Bob Fortenberry
9/28/86

Telecommunications in the Post-Divestiture Era

Telecommunications in the Post-Divestiture Era

Essays in Honor of Jasper N. Dorsey and Ben T. Wiggins

Edited by
Albert L. Danielsen
David R. Kamerschen
The University of Georgia

Lexington Books
D.C. Heath and Company/Lexington, Massachusetts/Toronto

Library of Congress Cataloging-in-Publication Data

Telecommunications in the post-divestiture era.

Includes index.
1. Telecommunication—United States. 2. Telecommunication—Law and legislation—Economic aspects—United States. 3. American Telephone and Telegraph Company—Reorganization. 4. Wiggins, Ben T. 5. Dorsey, Jasper. I. Wiggins, Ben T. II. Dorsey, Jasper. III. Danielsen, Albert L. IV. Kamerschen, David R.
HE7775.T355 1986 384′.0973 86–45505
ISBN 0–669–13445–7 (alk. paper)

Published simultaneously in Canada
Printed in the United States of America
International Standard Book Number: 0–669–13445–7
Library of Congress Catalog Card Number 86–45505

The paper used in this publication meets the minimum requirements of Amercian National Standard for Information Sciences—Permanence of Paper for Printed Library Materials, ANSI Z39.48–1984. ∞™

86 87 88 89 90 8 7 6 5 4 3 2 1

Contents

Preface

The telecommunications industry has recently undergone such dramatic changes that many no longer regard it as a public utility. Therefore, it is appropriate for this book to be entitled *Telecommunications in the Post-Divestiture Era*. Some day, economic historians and industry analysts will doubtlessly look on the AT&T divestiture as a watershed between a highly regulated public utility and a more competitive market environment.

The editors of this book are involved heavily in the ongoing dialogue regarding the telecommunications industry. In 1981, we developed the annual Public Utilities Conference at the University of Georgia. It draws participants from the electric, gas, and telecommunications industries. A total of five conferences have been held since then. However, the level of participation and interest in telecommunications has increased dramatically since the first conference—especially since divestiture became an imminent reality.

The driving force of technological change is behind many industry changes. Many of the issues of the pre-divestiture era remain the same (the extent of cross-subsidization, lifeline rates, and so on). However, the terminology and jargon used to describe and analyze these phenomena have been altered—for example, from "separations and settlements" to "access." There has been an increased level of interest in the telecommunications industry, as evidenced by the dramatic rise in participation in our conference by the new ATT Communications, the regional Bell Operating Companies, independents, and newcomers to the industry. Participation has increased more than fourfold since the inception of the conference. Now, nearly one hundred telecommunications industry and public service commissioners attend. Telecommunications has indeed become a hot issue.

This book reflects the basic philosophy of our academic and professional interests. Broadly educational, it is designed to stimulate the interest of middle and top management in telecommunications companies, consumer and commercial interest organizations, public service commissions, consulting firms, and academic institutions. The basic purpose is to promote an exchange of ideas among individuals involved in the telecommunications industries.

The focus of part I, "The Regulator's Perspective," is on state regulators and the decision-making process at the state level. It provides a view of telecommunications regulation, deregulation, and divestiture, from the point of view of those who make decisions in rate cases at the state level. It features presentations by Commissioners Henry Yonce (South Carolina), A. Hartwell Campbell (North Carolina), and Neilsen Cochran (Mississippi). In addition, there is a chapter by staff members of a prestigious public service commission—J. Robert Malko and Gregory Enholm (Wisconsin). Finally, a group of academicians—Homayoun Hajiran, David R. Kamerschen, and John Legler—discuss the economic and political determinants of the rates of return granted by the state regulators.

Part II, which is devoted to six presentations by professionals closely associated with telecommunications companies, is entitled "The Companies' Perspective." We have tried to strike a balance between presentations by professionals from the major interstate carriers and those from the local distribution companies. The purpose is to present the telecommunications issues from the standpoint of those who deal with them on a day-to-day basis. The principal issues here are access charges from the standpoint of the dominant carrier (Roy Billinghurst, AT&T), from the standpoint of the other interexchange carriers (Ferguson, MCI), and from the standpoint of the local-exchange carriers (Price, BellSouth). In addition, the related issue of bypass is discussed (Weber, AT&T). The agenda of unfinished business in the telecommunications industry from the perspective of a company president (Skinner, Southern Bell) and another view on the transitions from the standpoint of the local-exchange carriers (Ward H. White, USTA) are also presented.

Parts III and IV are devoted to some of the major issues in the telecommunication industry in the mid-1980s. In particular, part III is devoted to the pricing of telecommunications in an asymmetrically regulated market. Again, the purpose here is to present a variety of points by individuals who are on the firing line daily and who are associated with the major interexchange and inter-LATA carriers. Part IV, on the other hand, deals with nonprice telecommunications issues, such as depreciation (Terence Robinson) and interexchange issues (Robert Fortenberry), as well as with modeling state-level economic impacts of changing access charges (R. Carter Hill, Albert L. Danielsen, and David R. Kamerschen). It also provides a comparison of the telecommunications and natural gas industries (William Hederman), showing that many of the same basic issues are applicable to both industries and that each is in a state of transition.

This book is dedicated to two individuals who have influenced the theory and practice of regulation of public utilities in Georgia. We have provided in the next section detailed biographies of these distinguished doyens, who were honorees at previous University of Georgia Public Utilities Conferences (1982 and 1983). Jasper Dorsey, retired former vice-President and chief executive officer of Georgia Operations (Southern Bell), has been a long-time supporter of the University of Georgia and of our programs in public utilities there. Ben T. Wiggins, presently in private law practice and a former public service commissioner in Georgia for twenty-two years, is a personal and professional friend and a regular participant at our annual University of Georgia Public Utilities Conference.

Biographies of Honorees

Jasper N. Dorsey

Jasper Dorsey is a syndicated columnist, adjunct professor of management at the University of Georgia College of Business Administration, and contributing editor of *SKY* magazine. A native of Marietta, Mr. Dorsey attended the University of Georgia, earning his degree in journalism in 1936. He then enrolled in the University of Georgia Lumpkin School of Law and joined Southern Bell in 1938. During his distinguished career with Southern Bell and AT&T, he served in various executive capacities in Miami, Jacksonville, New Orleans, and Washington, D.C. From 1968 until he "retired" in 1978, he served as vice-president and chief executive officer of Georgia Operations. Prior to that, he was manager of governmental relations for AT&T in Washington. A man of immense energy, Jasper Dorsey remains a leader in business, government, education, and his church.

Currently, Mr. Dorsey maintains directorships of the Bank of the South, Bank South Corporation, and Research Atlanta. He has also been appointed to a fourth term on the Georgia World Congress Center Authority. Culminating many years of active involvement with the Georgia Chamber of Commerce, he has been named director for life. His distinguished academic and professional life gained him membership to the Blue Key National Honor Fraternity, ODK National Leadership Fraternity, SPHINX Honor Society, Gridiron, Beta Gamma Sigma and Phi Kappa Phi National Scholastic Honor Fraternities, Greek Horsemen, and Sigma Delta Chi Society for Professional Journalists. Mr. Dorsey continues his long-standing relationship with the University of Georgia as chairman of the board of trustees of the University of Georgia Foundation, president and chairman of the board of the University of Georgia Alumni Society, and charter member of the advisory board of the College of Business Administration. In addition, he was instrumental in founding the advisory board of the Henry Grady School of Journalism at the University of Georgia and was its first president.

Mr. Dorsey has generously served his community and the state of Georgia. Presently, he is on the boards of the Salvation Army, the Georgia 4-H Foundation, and the Georgia Trust for Historic Preservation. He has also chaired fund drives for the Atlanta Jaycees, the Atlanta Journal Empty Stocking Fund, the Georgia Foundation for Independent Colleges, the Georgia Trust for Historic Preservation, the State YMCA of Georgia, Junior Achievement, and Ducks Unlimited.

Ben T. Wiggins

Ben T. Wiggins has been a highly respected leader in public utility regulation for more than twenty years. His impressive record of service began when he became a member of the Georgia Public Service Commission in April 1957. He served as vice-chairman of the commission from 1965 to 1970 and as chairman from 1971 to 1978. During this time, he also held positions of leadership in the National Association of Regulatory Utility Commissioners (NARUC). He served on the Executive Committee from 1969 to 1978 and held the office of president in 1973 and 1974. His peers honored him for his outstanding contributions by naming him president of the NARUC Commissioners Emeritus Association in 1981.

A member of the State Bar of Georgia, Mr. Wiggins received his LL.B. from Atlanta Law School following undergraduate work at the University of Georgia, Atlanta. After a tour of duty with the 15th Air Force, he set up law practice in Toccoa, Georgia, in June 1946.

In 1951, Mr. Wiggins represented Stephens County in the Georgia House of Representatives. He took time off from his law practice in 1954 to become executive secretary to the governor of Georgia.

Mr. Wiggins has been actively involved in guiding the growth of his home state by serving as chairman of both the Jekyll Island State Park Authority (1972–77) and the Planning Committee of the Stone Mountain Memorial Association (1973–78). In 1979, he reestablished his practice of law in Atlanta, where he remains active in civic affairs through the Atlanta Lions Club and the Georgia chapter of the Leukemia Society of America.

Part I
The Regulator's Perspective

Part 1 consists of five chapters on the process of regulation and how it is viewed by state regulators. The first three chapters were written by prominent state regulators who are on the firing line in state and federal-state post-divestiture issues. Commissioner Henry G. Yonce (South Carolina) focuses on the issue of equal access and its effects on the activities of regulatory commissions, commission staff, the local-exchange carriers, and consumers. He warns that all market participants need to understand the issue of equal access, that the problem is a transitory one as we move toward a more competitive market, and that new problems will emerge as the equal access issue is resolved successfully. Regulation and deregulation are ongoing processes that will continue, it is hoped, with an end result that is beneficial to all market participants.

Commissioner A. Hartwell Campbell (North Carolina) questions the assumption that regulation is a necessary evil compared to competition. Even if competition were possible in the real world—and it is not in all instances—externalities (for example, public safety and pollution controls) and considerations of equity make regulation not only a necessary but a desirable adjunct to competition. Commissioner Campbell subscribes to a life-cycle model and provides numerous examples of how technology leads to the development of new industries, growth, maturation, and change. In his view, each stage of industry development offers new challenges and opportunities to regulators. He believes that the present era is a challenging one and that more flexible regulations and more responsive regulators are required to cope with rapidly changing conditions.

The third chapter, by Neilsen Cochran (Mississippi), provides insights into how a newly elected commissioner perceives the AT&T divestiture. His opinion is that the primary reason that national surveys have shown overwhelming disquietude with telecommunications service is that everyone was ill-prepared for such a watershed event. He illustrates his major point with a brief review of how lifeline telephone service and capital recovery are now being handled. This former professional baseball player is as serious about competing vigorously to establish a viable communication system as he was on the baseball diamond.

The fourth chapter was written by staff experts from the Wisconsin Public Service Commission, which has long been a leader in regulatory analysis. In their paper, Dr. Gregory B. Enholm and Dr. J. Robert Malko investigate what determines regulatory decisions on the cost of equity capital. They use the newly established subsidiaries of Ameritech as a case study. They first provide a history of divestiture and reorganization and outline the scope of operations for the seven regional holding companies. They also explain the difficulty regulators face in deciding economic and financial issues when they frequently have very little expertise in accounting, economics, or finance. Regulators often rely on the judgment of experts, but there are typically at least three points of view: that of the applicant, the commission staff, and intervenors. Enholm and Malko find that in the five Ameritech cases they studied in (Illinois, Ohio, Indiana, Michigan, and Wisconsin) during 1983 and 1984, regulators were confronted with a variety of methodologies (formal and informal) and comparison groups. However, staff witnesses appeared to influence the regulators more than did company or intervenor witnesses. The methodology underlying this assessment is judgmental rather than statistical.

The chapter by Homayoun Hajiran, David R. Kamerschen, and John B. Legler is designed to model what factors influence (1) a regulated firm's decision to file for a rate hearing; (2) the rate of return a regulated firm is likely to request; and (3) the probable rate a regulated firm will be granted by the regulating commission in a formal hearing. Their study is based on the regulation of public utilities by the Georgia Public Service Commission, which is unusual in that it is elected by the voters. Telephone, gas, and electric company data are used in the study, but primary attention is given the electric utility industry, since (1) more complete and consistent data are available for electric companies; (2) it is the utility most commonly considered in previous studies; (3) it is the only type of utility used for estimating one of the three equations in all previous modeling efforts reviewed. However, the authors believe that many of their conclusions are robust over all types of utilities. In general, the authors found that political factors are more prevalent in an elected-commissioner state in a regulated utility's decision to file. However, the estimated results of their last two decision equations, were similar in elected-commissioner and appointed-commissioner states in that the utilities' requested rate of return and the commissioners' permitted or allowed rate of return are apparently influenced by a common set of variables across states.

1
Regulating the Telecommunication "Bear" Markets

Henry G. Yonce
South Carolina Public Service Commission

In recent years, the approach to telecommunications issues adopted by state regulators and the Federal Communications Commission (FCC) has sounded very much like the deal made in an old tale by two friends on a hunting trip: Two friends went hunting, and as they sat around the hunting cabin the night before the big hunt, they decided to make this hunt competitive. So they placed a little wager on it. They agreed that when one of them brought something in—whatever it might be—the other one had to clean it. The next morning, one of the men was awakened by someone shouting. He jumped up and realized that it was his friend hollering "Open the door! Open the Door!" He ran to the cabin door and threw it open. Down the path came his friend, running as hard as he could, and right behind him came a big black bear. The friend ran through the cabin—with the bear still behind him—and out the back door, closing the door behind him and leaving the bear in the cabin. On his way out, he shouted, "Okay, I caught this one. You skin him; I'll be back with another one in a few minutes." State regulators have been seeing a lot of "bears" that have been chased into the cabin by the FCC and the courts and left for us to handle.

One of the most current "bears" we have faced is equal access. Not only state regulators but also telephone utility executives and consumers should be concerned and should be knowledgeable about the subject—for two primary reasons.

First, newspapers across the country, big and small, have carried a lot of publicity about the recent opening of equal access in Charleston, West Virginia. It has been a big media event. Customers have been—in their terms—badgered or harassed by the various carriers as they try to get their long-distance business, and the overall feeling is that confusion reigns. When an event such as this (a result of regulatory action, by the FCC, by the courts, or by a state commission) becomes a media event, it involves everyone. State regulators must be in a position to understand what is happening and to explain it to consumers, because the public expects regulators to know what is going on. Telephone executives must do the same thing. Consumers need to know the facts—to be able to weigh what is being offered, cut through all of the publicity, and make good intelligent decisions on what their options are.

In addition, one of the requirements of equal access is that all costs of equal access be identified separately from the operating companies' investment to ensure that equal access costs are recovered from access charges and not included in the rate base for subscribers. Therefore, it is incumbent upon consumers and regulators to ensure that these costs are indeed separated from the companies' expenses and investments and are recovered in the appropriate tariff.

Equal Access: How It Came About, What It Is, and How It Affects Consumers

Under the terms of the Modified Final Judgment, which brought about divestiture, the Bell Operating Companies (BOCs) are required to equip all conforming central offices to provide equal access by September 1986. There is a similar provision for the General Telephone Company and United; however, other independent telephone companies do not have this requirement. Equal access will provide the customer with the same type of access to all subscribing common carrier long-distance networks. In other words, the customer can dial the same number of digits and receive the same quality of service to complete a call from point A to point B, regardless of the carrier used. In an equal access-equipped office, the customers have the option of selecting in advance or of presubscribing to any long-distance carrier serving that office; however, they can presubscribe to only one carrier.

If they do presubscribe, all 1+ or 0+ long-distance calls that are (at least in South Carolina's case) inter-LATA, (local access and transportation) calls would go to that carrier—be it AT&T, MCI, GTE-Sprint, USTS, or any other.

In addition, all customers in an equal access office can access any long-distance carrier serving that office by dialing a five-digit code—10XXX. It is expected, however, that only those customers who are subscribers of that long-distance carrier and who have personal identification numbers (or whatever is required by that carrier) can use that five-digit code. Also under the terms of the Modified Final Judgment, the Bell Operating Companies (BOCs) have a responsibility to inform customers about their options under equal access to ensure that all customers know what is happening, what they as customers can do, and what they cannot do.

At least ninety days prior to conversion of a central office to equal access, the Bell Operating Company must inform each customer of the options and must provide a reasonable amount of information on the various carriers that have indicated a desire to be included in the notification. The information must be provided in a format that ensures that no one carrier is given an unfair advantage over others. In other words, when a list of carriers is provided to the customers, the names of the carriers must be randomly distributed so that no single carrier will be first on any list provided to all customers in that particular exchange.

During the ninety days following conversion to equal access, Southern Bell in South Carolina will also provide an informational program that reminds customers of their options and of the fact that they can choose and presubscribe to any of the carriers serving the office.

Under Southern Bell's proposal in South Carolina, there will be no service charge for choosing a preferred carrier, so long as the selection is made within six months after conversion to equal access. There will be a service charge for subsequent changes or for changes that are made after the six-month period. Southern Bell and most of the other BOCs will default to AT&T (that is, route all inter-LATA calls to AT&T) for those customers who do not select a preferred carrier. This arrangement prevents costly network rearrangements, since the calls are currently routed in that manner. However, new customers who apply for service after an office has been converted to equal access must select a carrier. If they do not, they will not be able to dial long-distance calls. Instead they will reach a recording that will tell them that they must dial their chosen carrier. Of course, if they do not select a preferred carrier, they will still have the option of dialing the five-digit code and reaching any of the carriers with which they subscribe.

The Bell Operating Companies also have a responsibility with regard to the carriers. Six months before the central office conversion, all carriers that are providing service in the state must be notified of the availability date of equal access in that particular office.

Four months before conversion, the carrier must notify the Bell Operating Company if it wants to be included in the customer information packet that is sent to all customers. The letter the carrier submits to the Bell Operating Company must specify the end office or offices into which that carrier wishes to provide service. The Bell Operating Companies will also accept from each carrier information concerning services offered, pricing structure, special rate offerings, monthly minimums, or any other information that the carrier desires placed in the package, including how a customer can apply for service.

Two weeks after the information is received from the carriers, as a service, the Bell Operating Companies will provide to all carriers included in the package a list of all other carriers in that package. When they notify the carriers of the equal access conversion of an end office, BOCs must also provide those cariers with a list of customers in that office. In addition, the Bell Operating Companies will also provide to the carriers—at their request and for a fee—lists of customers that are not presubscribed and those that are presubscribed to that particular carrier. As part of the presubscription process, carriers may act as agents on behalf of their customers to presubscribe customers.

As stated earlier, one of the other requirements placed on the Bell Operating Companies is the identification and documentation of the costs of equal access for the various exchanges. The BOCs must identify separately the costs of equal access on a central office by central office basis, and ensure that these costs are not included in any of the expense and investment data that is furnished for intrastate local-exchange rate making. Revenues to recover the intrastate portion of these costs must be requested separately from the state commissions.

What Lies Ahead

The equal access issue is just heating up and is going to get hotter as all of the various operating companies begin the customer and carrier notification processes. To meet the equal access requirements of the Modified Final Judgment, which must be completed by September 1986, all local Bell Operating Companies must begin converting these offices. As soon as those notifications and conversion processes have started, customers, utility executives, and state regulators will be affected and will become an integral part of the process.

Therefore, it is incumbent upon all of us to understand all the details about equal access—to understand the whys and the wherefores—so that we all can handle this "bear" in the easiest way possible. Of course, when it has been handled, we will have to stand around and wait for the next "bear" to come through—and goodness knows what that might be.

2
The Changing World of Regulation

A. Hartwell Campbell
North Carolina Utilities Commission

It has been said that the only thing that does not change is change itself. Certainly life itself is a changing process, and all of us are caught up in the very dynamic changes that occur daily. That is why we listen to the latest news—to be able to take note of what is changing now.

The world of business is one of constant change. Those of us who are related to the world of business as regulators find ourselves facing a changing situation. In fact, the very process of regulation calls for a continuing review of changes—involving the quantity and quality of service, sources of supply, costs of operations, and returns on equity consistent with the established laws and regulation as given to us, and as may be tempered by evidence and judgments. This is commonly known as rate-base regulation. And this is the consistent business of regulation. Too often, we become so engrossed with the process of regulation that we fail to see the broader implications of just what is happening to regulation.

It has often been assumed that regulation is a necessary evil, that it must replace competition to achieve adequate nondiscriminatory service of an essential utility at prices deemed just and reasonable to customers and reasonably profitable to the owners.

The implication has always been that regulation is a proxy for competition and that even when it functions properly, it is a second-best effort. Pure economic theory seems to say that if the free marketplace were allowed to operate, competition would allocate the limited resources and services in the most efficient manner. However, the real world is such that we do not have pure competition, and we never shall. Furthermore, societal needs are such that from time to time, we all want to interrupt the most efficient manner of allocating resources in favor of other concerns, such as public safety, pollution controls, and personal concerns for the economically deprived members of society.

This is just another way of reminding ourselves that even if we had a free marketplace with the law of supply and demand used to its utmost efficiency, there would be a place for regulation that would rise above pure economic concerns. Having recognized that, let us put aside these other functions of regulation and focus more specifically on the changes in regulation that do deal with economics. Assume, for the purposes of our thesis, that regulation is the proxy for competition. Assume further that competition is becoming more and more of a factor in our so-called monopolistic utility services. The question then becomes how regulation will deal with the emerging growth of competition in regulated industries.

It may be of some worth to examine why competition is on the move in our economy. To accomplish this, it may profit us to look into the life cycle of industries, both unregulated and regulated.

Katherin Miller of the Electric Power Research Institute (EPRI) presented an interesting essay in the June 15, 1985, edition of *Fortnightly,* in which she postulates that all of us have paid too little attention to the life-cycle theory of business enterprises. She concludes that there are four distinct and discernible stages in the life of a business. Stage one is the discovery or introduction of a business or service, such as the telephone. When Alexander Graham Bell first said, "Watson, come here," it was the beginning of something that had no known form or usages. It was a novel curiosity. After the telephone was patented, the owners of the patent offered to sell the whole process to Western Union for the sum of $100,000, but Western Union foolishly turned it down.

Another interesting story on beginnings is about a former iron miner named Carl Wickman who sank all of his savings into a Hupmobile agency in Hibbing, Minnesota. In 1914, after failing to find a buyer for his inventory, which consisted of one lone Hupmobile, Wickman squeezed ten seats into the seven-seat touring car and began making hourly trips from a saloon in Hibbing to a town named Alice, some four miles away. By 1925, Wickman's tiny company was sold, and its new owners called it the Greyhound Lines.

In the early and introductory days of these new enterprises, the need for regulation was not recognized. However, that need soon became obvious. In my home community of Raleigh, North Carolina, I have been told that there were once two telephone companies, one with wires on one side of the street and the other with wires on the opposite side.

Ms. Miller says stage two of any new utility is one of rapid growth, during which new markets and application appear. Expanded acceptance is accompanied by enlarged capacity. This develops into economies of scale, which bring on declining prices. Declining prices provide for further accelerated growth.

Stage three is that of a mature industry approaching market saturation. New technologies are accommodated without major obsolescence. However, after the rapid growth and the stability of maturity have been assured, false assumptions are too often made. Both the electric utilities and regulators accepted the annual growth of 7 to 8 percent as something that was assured. Production plants were planned and begun with that assumption. Other economic forecasts were not properly factored into plant planning. The result has been a temporary overcapacity, bringing about a rate base that is burdensome to consumers and embarrassing to the planners.

Eventually, stage four appears when changing conditions cause pressures and new competition to the extent that mergers, consolidations, and even bankruptcies occur. Some utilities are in this consolidation or decline period today.

To properly illustrate the life-cycle theory of a utility, we have to look no further than the railway system in America. The introductory period, from 1820 to 1835, was followed by a period of rapid growth, from 1835 to 1910. Maturity arrived and lasted through 1960. For the railroads, 1960 marked the point at which competition from the airlines signalled the end of passenger service. During this stage, many major railroads went into bankruptcy. Mergers and nationalization resulted. The number of Class 1 roads has been reduced from a high of 106 to 22, and track mileage has shrunk from 350,000 to 180,000 today. It is true that railroad freight tonnage has shown some growth; nevertheless, the market share of tonnage has declined.

This fourth stage of the life cycle may be appearing in utility after utility. In communications, we have seen the breakup of the greatest business corporation in the world. We have competing carriers, cellular telephones, space satellites, legal competition, and a whole new situation. Natural gas has certainly entered a different form, with strong competition from other energy sources. Buses are all but deregulated, certainly in an intrastate jurisdiction. Airlines have escaped deregulation altogether so far as entry, routes, schedules and prices are concerned. Rail passenger service has been nationalized.

If business enterprises that are regulated have a business cycle, it goes without saying that the business of regulation also has changed in remarkable ways. In cycle one, regulation is hardly present, if at all. In rapid growth periods for industries, regulation sees rapid growth. When rapid growth brings economies of scale, it is easy to be a regulator. About the only need for rate proceedings is to reduce prices to reflect reduced costs. Until the mid-1960s, there were few or no consumer revolts with which to deal, but in stages three and four, regulation has been tough. Competition has caused regulators to ponder the future role of regulation. Not many agencies will meet the fate of the Civil Aeronautics Board and find themselves dissolved, but it does bring on a time of assessment and reevaluation for regulators.

Therefore, let us look at competition and regulation together for a moment. Just as there has never been a complete monopoly that escaped indirect competition in the past, it is unlikely that competition will be pure competition to the degree that rate-base prices can be avoided. One of the key differences between regulated and competitive prices is that with competition, the prices that all companies use in a particular market will all be driven to the same price. Not so with regulation, in which the prices established are designed to return all true costs to a company, including an attempt to earn a rate of return on embedded assets. Competitive prices cannot be established on a theory of the recovery of all costs plus a return on equity.

Another remarkable problem with competitive pricing versus regulatory pricing is that competitive forces cannot and will not fit into a "test year" concept, and, certainly, competitive pricing will not tolerate a regulatory lag. The basic problem is that competitive firms cut prices when demand is slack and raise them when demand is high. Furthermore, pricing becomes a life-and-death issue of more immediacy with competitive firms.

From the standpoint of the investor in regulated companies, rates do not track prices that a competitive market would charge. Also, most often competition will prevent full recovery of allowed returns when competitive prices are lower, and, in return, regulation will prevent regulated companies from compensating for differences when higher prices can be tolerated. All of this can result in a death spiral for regulated services.

It seems obvious that there are many other factors that must be considered by regulators as we, too, are entering a strange new breed of competitive pressures. One could say that the doctrinaire model of regulation worked pretty well for a rather extended period. At least rate-base economics has worked for decades within acceptable tolerance levels of the using and consuming public and the regulated utilities. This tolerable balance is now being tested as never before.

So long as the regulator could order "toll booths" in the highways of pricing and all traffic had to pay the tariff, things were manageable. Once technology and competition made it possible for the largest customers of utilities to go around the turnstyle, what we now call a "bypass" began. Incidentally, the bypass is not something known only to the communications world.

If we really wish to know some of the consequences of competition on regulation, we can look to the airline industry, in which rate averaging is almost entirely gone. Deaveraging is the pricing mechanism in wide use. As surely as we have competitive common carriers in communications, we are going to have them going after the busy-circuit paths and disregarding the hinterlands. This will sooner or later bring about the deaveraging of rates. Cross-subsidization of a class of customers from the revenues of another will soon be an intolerable burden that cannot continue.

Regulators are seeing increasing conflict among classes of ratepayers in determining the proper sharing of cost responsibilities. Indeed, we are seeing some strange, paradoxical situations. In natural gas rates, we have to accommodate for industrial competition with flexible rates in order to better protect the residential customers from the responsibilities of assuming the larger share of fixed costs.

In summation, it seems obvious that we as regulators must face some major changes in the near future. We need to expand our economic understanding of the competitive forces in the economy, particularly as they move into contact with rate-base pricing. All of us could expand our understanding of the degree to which accommodation can be made to the expanding influence of competition on regulated industries.

We must consider and seek legislative changes that will enable us to find better ways to reduce regulatory lag, provide for more flexible pricing, and ensure an adequate supply of service with equity. Furthermore, we are going to have to reckon with the world of deaveraging flat rates while trying to avoid discrimination. Again, it is my belief that we must move from cost-based pricing to some value-of-service price recognition based on how the marketplace values a commodity or service. This may possibly call for the pricing of some services above costs to compensate for the pricing of other services below costs.

However, as we face this new era of regulation, one thing must be kept in mind. If rate-base regulatory pricing has been considered only second best at times, we need to recognize that competitive pricing is not perfect either. We only have to listen to the cries and screams coming from the jungle out there to appreciate that at times the competitive world may be less acceptable than regulation. Ask the shoe industry—or try textiles, furniture, automobile, or steel manufacturers. You can hear them all the way to Washington as they ask for protectionism—and what is protectionism but another crude form of regulation? Say all that you can about economic theory and the allocation of resources in our world, we all believe in free trade and competition—for the other fellow.

It is indeed my judgment that we may well be entering stage four in the profession of regulation—a stage that is going to be more challenging, more flexible, and more responsive to changing conditions that we have ever known. The old attitude—"Well, that's not the way we do things here"—has quietly gotten up and gone. The future of regulation is going to be a new and exciting place to be for people willing to accept change.

3
Divestiture: A Public Service Commissioner's Perspective

Neilsen Cochran
Mississippi Public Service Commission

The breakup of AT&T certainly has been a challenge to me as a newly elected public service commissioner. It has required some innovative approaches in an environment of confusion and frustration for both regulators and providers of services.

The preservation of local exchange service is my number-one priority, and this newly created environment must be treated as cautiously as one would handle a loaded gun. Some benefits may have already been derived from that divestiture, and possibly there will be more to come, but I would like to relate some of the more interesting and challenging issues that have arisen with regard to preservation of local service.

A recent survey of 1,700 companies by the National Conference Board indicates widespread displeasure with the situation as a whole. The survey included companies with sales greater than $10 million and the top 500 companies in sales, profits, employees, and market value as ranked by *Forbes* magazine. An overwhelming proportion (92 percent) of the corporations studied planned some sort of major telecommunications change within the next three years. Of these, 62 percent planned to switch to another carrier for their long-distance service. Over four-fifths of those surveyed reported that they believe service has deteriorated. The complaint most widely mentioned was difficulty in locating the personnel responsible for maintenance and information. Although the results of this survey are not completely representative of the situation that exists in Mississippi, the frustration exhibited by the people surveyed is somewhat universal, extending all the way to the regulator.

I feel that a majority of the commissioners and staff throughout the country were ill-prepared for the divestiture of AT&T. Numerous questions and issues immediately surfaced, and regulators, in my opinion, attempted to resolve most of these newly created competitive issues as if they were using a Band-Aid after the major surgery that was forced on consumers by the breakup of AT&T. There were numerous orders issued that were later amended or rescinded as the various effects of these orders became evident.

That frustration still exists as we all adjust and readjust to directives from the FCC and various courts of law.

The current topic of national interest and discussion is lifeline telephone service. In the wake of divestiture, predictions of skyrocketing local rates and tremendous customer dropoffs have brought this issue to the front lines of the regulatory battlefield. Members of Congress have introduced legislation that would provide for lifeline service to those in need. I believe that some of the best intentions in this area are misguided, and I am also convinced that there is a solution to the problem that would allow us to preserve the concept of universal service without penalizing taxpayers or telephone ratepayers.

First, a common definition of exactly what constitutes lifeline service should be determined by the states through our organization, the National Association of Regulatory Utility Commissioners (NARUC). When we are developing this definition, I believe we should consider lifeline as a generic or "no frills" type of service that would guarantee the end user the ability to access emergency numbers and to receive incoming calls. To me, this is what lifeline service is about—the ability to reach the outside world in times of need and the assurance that a concerned neighbor or relative can communicate with other neighbors and relatives. By no means should lifeline be considered a premium unlimited service, because that would result in the need for a subsidy from taxpayers. We must satisfy needs in every case only when it is economically feasible and without placing additional financial burdens on existing users of the services.

I believe that industry and regulators, collectively, can develop a program of lifeline telephone service that will meet the needs of the general public without putting an unnecessary burden on others. I feel that we should pursue a program of this nature, because regulators have an obligation to see to it that every modern convenience utilities can provide is made available to each and every citizen whenever possible. Communications with police, fire, and medical personnel are essential in this modern world. Perhaps the answer is a low fixed monthly charge, with a high per-minute message rate on local calls. In the event of an emergency, advance switching equipment could be programmed to allow free calls into emergency numbers. There are many alternatives to subsidized, premium, unlimited local service as a local solution to the lifeline problem that should be pursued in an effort to safeguard the interests of all ratepayers.

Another area of concern for me that has heated up recently and that promises to be one of the most critical areas facing regulators in the coming months is the issue of capital recovery or depreciation. Unless this issue is properly addressed, it promises to choke the remaining life out of our local-exchange companies. As technology speeds along and competition is introduced to more and more historically monopolisitic areas, underrecovered capital becomes a greater problem.

In the survey by the National Conference Board, mentioned earlier, over two-thirds of the respondents expressed significant interest in using bypass techniques. A quarter of the firms already do bypass, and the remainder of the firms are involved, to varying degrees, in evaluating the process. Three-quarters of the firms plan installation no later than 1987, and another 13 percent expect to use bypass techniques after 1988. Bypass is an issue with which we have not had to deal with on a large scale so far in Mississippi. The most active area of bypass in Mississippi appears to be tariff shopping, where the local companies' own tariffs have been used to obtain private lines that allow direct links with carriers, which, in turn, can avoid paying for access. We are aware that bypass is an immediate threat to local service. We must proceed to look at the inequity of intrastate access charges, and we must begin to deaverage prices. But all of this should be done in a controlled process, with the best interests of all users at heart. It is virtually impossible to attempt to regulate technology—and quite frankly, it would be stupid to do so—but there has to be some middle ground that can be reached, leaving our communication system intact with numerous choices for customers, not only in metropolitan areas but also in rural areas.

I would like to think that I have shown you some of the problems that we face as regulators. I am certain that our situation in Mississippi is not identical to that of any other state, although it may parallel some rather closely. We do want a manageable form of competition. We definitely want to preserve a viable communication system for all users. And daily we are being educated about means with which to operate in this divested environment. Surely, as we learn, we will progress.

4
Cost-of-Equity Decisions by State Regulators of Ameritech's Subsidiaries

Gregory B. Enholm
Salomon Brothers, Inc.
J. Robert Malko
Public Service Commission of Wisconsin

The telephone industry in the United States has been dominated by American Telephone and Telegraph Company (AT&T) since its inception approximately 100 years ago. During the past thirty years, technological development, erosion of entry barriers, and several legal challenges have changed the telecommunication industry and have threatened the ability of AT&T to maintain its dominant position. In 1982, as a result of the landmark settlement of an antitrust suit brought by the U.S. Justice Department, AT&T agreed to a substantial reorganization and divestiture of some significant operations, effective in January 1984. According to Charles L. Brown, chairman and chief executive officer of AT&T, "Traditionally we have priced our services to support the goal of universal service. No longer is this policy tenable. More and more we must relate rates to costs."[1]

This chapter is organized into the following primary sections: First, we provide a brief history of the breakup and reorganization of the Bell Telephone System. Second, we discuss the procedures used in determining rate cases. Third, we present an assessment of the basic assumptions, the methodology, and the impact of return-on-equity testimony. Fourth, we provide a summary of regulatory decisions in Ameritech rate cases. Fifth, we discuss the estimation of Ameritech's cost of common equity. Finally, we present conclusions and future directions for analysis.

The History of Divestiture and Reorganization

Since Alexander Graham Bell was granted a patent entitled "An Improvement in Telegraphy" in March 1876, the Bell Telephone System has pursued the objectives of building a national telephone system operated by one company.[2] The pursuit of this objective has resulted in several antitrust actions by the U.S. Justice Department and by competitors. The current restructuring of American Telephone and Telegraph Company is primarily a product of two separate U.S. Justice Department antitrust actions, filed during 1949 and 1974, which alleged violations under the Sherman Act. The first action, the Western Electric action, sought the divestiture of Western Electric and the nondiscriminatory licensing of AT&T patents. This dispute was settled during 1956 by a consent decree that (1) prohibited AT&T from engaging in any business other than telephone services, (2) generally restricted Western Electric to manufacturing for the Bell System, and (3) required that patent licenses be granted to all applicants.

The second action by the Justice Department sought a significant restructuring of AT&T that included the divestiture of the twenty-two telephone operating companies, Western Electric, and parts of Bell Laboratories. After several delays, the federal court approved and entered a consent decree, the Modification of Final Judgment, on August 24, 1982, and AT&T and the Justice Department agreed to this judgment. Federal Judge Harold Greene played a major role in these activities. The U.S. Supreme Court affirmed the Modification of Final Judgment on February 28, 1983.

Some important terms of the Modification of Final Judgment included the following:

1. AT&T must divest its local-exchange operations.
2. The local-exchange companies must provide equal access to all interexchange carriers.
3. The interexchange and terminal equipment portions of AT&T are free from the competitive restraints of the 1956 consent decree.
4. Local-exchange companies may not provide interexchange services or other services not within their natural monopolies.
5. AT&T must submit a plan of reorganization for approval by the U.S. Justice Department and by Judge Greene.

A significant result of the Modification of Final Judgment is the recognition of competition as a desirable and growing factor in the telecommunications industry. In the long run, all areas of the communications industry could support competition. However, only two of the former (pre-divestiture) Bell System's three major market areas—long-distance transmission and customer premises equipment—are no longer considered natural monopolies. The third market, local exchange, will continue to be operated as a regulated monopoly.

The reorganization plan from AT&T, as modified by Judge Greene, was agreed to during 1983 and implemented in January 1984. This plan attempts to meet the objective of competition (where feasible). The following are some important components of the implemented plan:

1. AT&T has been divested of its twenty-two telephone operating companies, and these operating companies have been organized into seven regional holding companies that are approximately equal in (financial) size.
2. AT&T retains the Long Lines Division, Western Electric, Bell Laboratories, and the International Division. Western Electric and Bell Laboratories are permitted to provide services for customers that are outside of the Bell companies. However, AT&T has lost its exclusive license contracts with the operating companies.
3. The divested local-exchange companies have been restricted to providing service within local access and transport areas (LATAs), and to providing local interconnection to long-distance carriers. These companies can sell customer equipment and license its manufacture by others, but they cannot manufacture customer equipment.

The following seven regional holding companies have been formed from the twenty-two Bell operating companies and are currently conducting activities and operations: (1) Ameritech, (2) Bell Atlantic, (3) Bell South, (4) NYNEX, (5) Pacific Telesis, (6) Southwestern Bell, and (7) U.S. West.[3]

American Information Technologies, Inc. (Ameritech), has combined and is composed of five operating companies covering five states in the Great Lakes region: Illinois, Indiana, Michigan, Ohio, and Wisconsin.

The parent-subsidiary organization of the regional holding companies is guided by the need to keep unregulated business activities separate from the local-exchange company rate base. Primary business activities of the regional holding companies include (1) provision of basic telephone services; (2) connection to interexchange companies; (3) publication of white and yellow pages and other directory services; (4) cellular mobile telephone service; and (5) new ventures through fully separate subsidiaries.

Regulatory Procedures in Rate Cases

Under current regulatory procedures, the decision of a utility to file a request for a rate increase with a regulatory commission is usually based on the utility's inability to earn the previously allowed return on common equity. Even small deviations from allowed return on equity result in changes in net income of $1 million or more.

Utilities are also very capital-intensive compared to most unregulated companies. The return on this capital, therefore, will be a major factor in the determination of any rate increase. When a rate increase request is filed, the company must present evidence to justify the need for a rate increase. Generally, commissions use historical cost, as well as accounting information on sales and operating expenses, to determine the carrying charge for mortgage debt and preferred stock. However, common equity does not have a historical cost; therefore, its cost must be estimated for each rate case. In estimating the cost of common equity, the commission will consider testimony from expert witnesses on what investors expect to earn on the utility's common equity. Witnesses attempt to estimate investors' expected return on equity by using cost-of-equity methodologies such as discounted cash flow, comparable earnings, the capital asset pricing model, and risk premium.[4] Interested parties, called intervenors, and commission staff are also permitted to file testimony on expected return on equity.

The commission then will deliberate to determine an allowed return on equity that is based on (but not always equal to) the expected return on equity. This allowed return on common equity is used to establish new rates. The company will then experience an earned return on equity based on the revenues earned and costs incurred with these new rates. If this earned return on common equity is inadequate, the utility will file another request for a rate increase.

Although this process can, in theory, be explained in a straightforward matter, commissions do not always find the process easy to follow in practice. The following quotation from the Opinion and Order of the Public Utilities Commission of Ohio in the Ohio Bell Telephone Company case number 83-300-TP-AIR gives an example of how difficult the process can be in practice for commissioners who generally are not experts in financial matters:

> In determining total allowable revenues for a utility company, the commission, of necessity, must make a large number of individual decisions with respect to specific issues. In making most of these decisions the commission is confronted with perhaps three or four recommendations or alternatives. However, with the selection of a fair and reasonable return on common equity, the commission is confronted with numerous theories and models as well as vast amounts of relevant data. Of course, the commission can only select one rate of return. To accomplish this, the commission must select the most appropriate method and the most relevant data and apply them to determine an appropriate rate of return. This does not mean that the commission rejects other data or other techniques as unacceptable, but merely that the commission exercises its judgment and selects that recommendation it believes to be the most reasonable given the facts and circumstances presented. (p. 49)

Cost-of-Capital Testimony

In 1983 or 1984, the five state commissions that regulate the operating companies of the new midwest Bell holding company, Ameritech, conducted rate increase cases. Return-on-equity testimony was filed in all five, with varying degrees of disagreement. In Illinois and Ohio, seven witnesses presented cost-of-equity testimony; in Indiana four did; in Michigan there were three; in Wisconsin, two. A total of twenty-three testimonies were presented by twenty-one witnesses (two witnesses presented testimony in two cases each).

Basic Assumptions

Even though the rate increase would take effect after divestiture, all but two witnesses based their cost-of-equity analysis on American Telephone and Telegraph. A study by Evans and Rothschild indicates that this approach could be appropriate, since modern finance theory maintains that value cannot be affected by repackaging the financial securities of a firm.[5]

Methodology

The twenty-three witnesses testifying in the five rate cases used four different methodologies in thirty-one different applications. Five witnesses did not use a formal methodology. The discounted cash flow (DCF) methodology was used by fifteen witnesses. The second most popular method was the comparable earnings methodology, which was used by nine witnesses. The capital asset pricing model (CAPM) was used by six witnesses. The risk premium methodology was used by two witnesses.

In applying the DCF methodology, the witnesses used a somewhat narrow range of forecasted dividend yields (7.5 percent to 10.47 percent). However, as is usually the case with the DCF methodology, dividend growth estimates varied widely (3.0 percent to 8.54 percent). Witnesses developed estimates of the expected return on common equity using the DCF methodology in a range of 10.5 percent to 17.86 percent.

There was no consistent use of the comparable earnings methodology by the witnesses. Each witness used different justifications for establishing comparable groups of companies. Overall, the witnesses found that the expected return on common equity using the comparable earnings methodology was 13 percent to 18.5 percent.

In applying the capital asset pricing model, the witnesses used a narrow range of risk-free returns (10.57 percent to 11.5 percent). The AT&T beta was estimated to be betwen .45 and .65. Witnesses disagreed, to a large extent, on the appropriate market risk premium. The lowest premium used was 3.7 percent and the highest was 8.9 percent. Applying the CAPM, the witnesses found estimates of the expected return on equity within the range of 13.55 percent to 16.95 percent. Two witnesses used a risk premium methodology. Their estimates of the risk-free return varied from 11.04 percent to 14 percent. The risk premium was estimated to be 3.66 percent to 5.5 percent. The witnesses estimated the cost of equity to be in the range of 15.48 percent to 17 percent, using the risk premium methodology.

Impact of Testimony

Staff witnesses appear to be the most successful in persuading the commissions to establish the allowed return on equity. In each case, the allowed return on equity was below the lowest expected return-on-equity estimates from the company witnesses. Intervenors were somewhat more successful than company witnesses. However, the commissions did not arrive at essentially identical estimates of the cost of equity for the five operating companies of AT&T (now Ameritech). The allowed returns on equity range from 13.83 percent to 15.85 percent.

The multihundred basis point ranges for each methodology served to reinforce the Public Utilities Commission of Ohio comments cited earlier. However, the results are not totally implausible in that the recommended ranges have a distinct pattern, determined by whether the witness is testifying for the company or for an intervenor or is a commission staff member.

For the average range of expected return on equity by type of witness (applicant, commission staff, or intervenor), the seven company witnesses presented the highest average range; six intervenor witnesses presented the lowest range; and three staff witnesses presented a range between the other two. Each group of witnesses used similar cost-of-equity methodologies in determining its ranges.

Commission Decisions in Ameritech Rate Cases

The Illinois Commerce Commission stated in Dockets 83–0005 and 83—0669:

> In reviewing the cost of capital to Illinois Bell Telephone and ATTIC post divestiture the commission is faced with a difficult task. The cost of capital cannot be determined exclusively through the use of models and formulae although they are helpful tools in a process which must involve an exercise of judgment by the commission. (p. 50)
>
> Although the testimony predicted dire effects of divestiture on AMERITECH, Mr. Schelke now concludes (in the 12/6/83 Smith Barney Report) that "for investors who feel compelled to concentrate holdings, we would suggest focusing on AMERITECH, Western Bell, and US West. . . ." In

> his specific discussion of AMERITECH, Mr. Schelke advises investors concerning "Ameritech's strong financial and cash flow position. The area's economy has been sharply depressed in recent years, particularly in Michigan and Ohio, but a pickup is likely as the economic recovery continues, with a subsequent favorable impact of demand for telecommunications services. We view AMERITECH as one of the stronger RHC's from both financial and operating standpoints." (p. 52)
>
> In light of these considerations and all the evidence of record in this proceeding, the commission is of the opinion that a fair rate of return on equity for Illinois Bell is that recommended by Dr. Bussa of the Commission's Economics and Rates Department, 15.85%. At the time of his testimony and cross examination, Dr. Bussa did not have the benefit of market trading in post divestiture AT&T shares and the federal court had not entered its order regarding ATTIC as the default carrier in Illinois. The commission considers these events to have a significant impact on the appropriate return on equity evaluation. (p. 52)

The Public Service Commission of Indiana stated in Docket 37200:

> In general, in this proceeding the commission is blessed with competent analysis of petitioner's cost for common equity by three experts. The evidence of record supports a cost for equity as low as 14.5% and as high as 18%. Having reviewed this evidence we conclude that an appropriate cost for an allowed return on common equity for petitioner is approximately 15.5%.
>
> Determination of the return allowed for common equity is perhaps the most judgmental determination which this commission is required to make in any general rate proceeding. In evaluating the testimony of Dr. Lewellen, Dr. Shearer and Mr. Lash we have, of course, attempted to strike an equitable balance between the interest of petitioners, ratepayers and the interest of petitioner's investors. We believe that our allowed return on common equity of 15.5% percent is well within the parameters of the evidence presented, will provide petitioner with the opportunity to attract capital at a reasonable rate and will provide petitioner's investors with the opportunity to earn a return commensurate with the risks attending investment and petitioner's divested operation. (pp. 24–25)

The parties in the Michigan Bell Telephone rate case number U-7473, agreed to a stipulation, in May 1983, regarding many aspects of the rate case, including the cost of equity. Only three witnesses presented cost-of-equity testimony. These three witnesses were testifying for Michigan Bell. The commission stipulation included an allowed return on common equity of 13.83 percent.

The Public Utilities Commission of Ohio stated in Docket 83–300–TP–AIR:

> After having reviewed the analyses and recommendations of all six witnesses who presented testimony, the commission is of the opinion that the recommendation of staff witness Wissman is the most reasonable presented. The staff's recommendation is based upon the DCF approach and looks at the most current available data for consolidating AT&T. The DCF methodology is consistent with economic efficiency because it equates the required return of the equity investor to what can be earned on a new additional investment in the competitive market place. Unlike the comparable earnings for the relative risk and earnings approach used by Mr. Langsam which looks at what has been earned on historical investments, the DCF looks at what can be earned on new investments which is consistent with the principals of economic efficiency. Therefore, we agree that the absence of data on the regional holding company makes it impossible to determine the current cost of equity for AMERITECH. But investors are and have been aware of divestiture and the use of current AT&T market data reflects the present as well as future expectations concerning divestiture. (p. 52)
>
> The commission, after considering all of the aforementioned factors, believes that the arguments for the low and high end of the range tend to balance out and has determined that the mid point of the staff's range, specifically 14.08%, represents a reasonable estimate of the cost of equity capital to the applicant. (p. 54)

The Public Service Commission of Wisconsin was more cryptic in its decision on the cost of common equity for Wisconsin Bell in Docket 6720–TR–36:

> Wisconsin Bell has requested a return of 16.7% on its own common equity which in this case would also equal a 16.7%return on AMERITECH's common equity invested in Wisconsin Bell. Staff testified that a return on the underlying common stock equity of Wisconsin Bell's parent of 15% would be reasonable at this time. The commission, after due consideration, finds a return of 14.25% on AMERITECH's common equity is reasonable and just for purposes of this proceeding. Since the percent of common equity for Wisconsin Bell and the consolidated AMERITECH are both estimated to be 60%, the 14.25% is also the return on the common equity of Wisconsin Bell. (p. 32)

Estimating Ameritech's Expected Return on Equity

When Ameritech's individual utility operating company subsidiaries file their next requests for rate increases with their respective state commissions, the testimony presented on Ameritech's cost of common equity will be based on Ameritech's own financial status. Several questions will arise regarding the estimation of an appropriate expected return on equity for Ameritech utility subsidiaries, including the following:

1. Will Ameritech's risk be higher, lower, or the same as the risk associated with the old Bell System? Finance theory indicates that, at least at the beginning, the risk should be the same as the risk associated with the old Bell System. However, when Ameritech establishes new nonutility subsidiaries, Ameritech's risk will likely be different from the risk associated with a pure Bell telephone company. If Ameritech chooses to enter businesses in which the risk is significantly different from the risk associated with a telephone utility, commissions will need to adjust any findings on Ameritech's cost of equity for the differential risks.
2. What companies will be comparable to Ameritech? So long as the other regional holding companies maintain a large percentage of their revenues and assets in telephone utilities, the obvious candidates for companies comparable to Ameritech are the other six Bell regional holding companies. It may also be the case that Ameritech will be comparable to independent telephone utilities.
3. Will the appropriateness of any cost-of-equity methodology be affected by the divestiture of the five midwestern operating companies from the Bell System? The primary difficulty that would be encountered by an expert witness attempting to use a cost-of-equity methodology to establish Ameritech's expected return on equity would be the difficulty of obtaining data on Ameritech.

In using the DCF cost-of-equity methodology, an expert witness should not encounter significant difficulties in estimating the dividend yield component. However, direct estimates of the dividend growth rate for Ameritech will be hampered by the lack of historical data on Ameritech dividends. It may be possible to substitute a dividend growth rate for the historical Bell System dividend growth or to use independent telephone company dividend growth data. Professor Charles Linke has been using a survey method to estimate what various financial analysts and institutional investors expect dividend growth rates to be for the Bell System.[6] With seven regional holding companies, such surveys may be of use in establishing general boundaries for dividend growth estimates.

The second most popular cost-of-equity methodology, comparable earnings, is not significantly affected by the divestiture of the five midwestern operating companies from the Bell System. Since comparable earnings is, to a large degree, based on historical data, this methodology should not be hampered by the lack of current data on Ameritech. However, the selection of comparable companies may become more difficult.

Using the CAPM cost-of-equity methodology on Ameritech will be complicated by the lack of data on which to calculate a beta for Ameritech. A company's beta is usually calculated using five years of historical beta. Again, it may be possible for expert witnesses to determine that the risk of Ameritech is not significantly different from the risk of the previous Bell System and, hence, it would still be appropriate to use a beta calculated on the basis of a historical Bell System beta.

Finally, an appropriate risk premium methodology could require data on Ameritech's common equity returns. However, as with CAPM, an expert witness may be able to determine that Ameritech's risk is sufficiently similar to the Bell System's historical risk that the Bell System data from previous years is still appropriate in this methodology.

Expert witnesses who have access to the appropriate data and are willing to attempt to compare groups of industries by comparable risk can attempt to estimate Ameritech's cost of equity by examining these other industries. Using a methodology such as tthe DCF, the expert witness could estimate the cost of equity for one set of companies and use that as an upper boundary, then use another set of companies as a lower boundary for the cost of equity. In that the breakup of the Bell System is a very unusual occurrence for the estimation of cost of equity, it will be interesting to see how expert witnesses deal with it.

Conclusions and Future Directions

With the divestiture of the operating companies from AT&T, it is likely that the financing of the new holding companies will become more complex. This likelihood and the upheavals in electric utility finance for nuclear power indicate a need for regulatory commissions to increase their capability to analyze utility financial issues.

Estimating the cost of common equity for the regional holding companies and their operating utilities is likely to be complicated, particularly if the holding companies engage in extensive nonutility activities or if the utility subsidiaries face different risk. Expert witnesses should reevaluate the appropriateness of their cost-of-equity methodologies in light of new conditions.

Allowed returns on common equity granted by the five commissions regulating Ameritech subsidiaries vary somewhat. It is important to note that prior to divestiture, AT&T was regulated by fifty state commissions. Ameritech is regulated by only five commissions. Therefore, Ameritech's financial condition may be more affected by individual regulatory decisions than was AT&T's.

In the future, regulatory commissions will face at least three challenges in regulating the financing of Ameritech utility subsidiaries. First, the estimation of the cost of common equity may be affected by the existence of the diversified holding company. Double leverage analysis may or may not be appropriate. Second, no historical data exist on Ameritech, making comparisons more difficult. Third, the mixture of competition and regulation is likely to create uncertainties for commissions, operating companies, and competitors. These three areas, as well as changes in other regulated areas, indicate that commissions may wish to consider regulatory strategic planning to address concerns about the future.[7]

Notes

1. Charles L. Brown, *Annual Report to Stockholders,* American Telephone and Telegraph Company, February 6, 1981.

2. Mark D. Luftig and Neil D. Yelsey, "American Telephone and Telegraph Company—The Bell No Longer Tolls," Salomon Brothers, Inc., Report, August 15, 1983. See also "Information Statement and Prospectus," prepared by American Telephone and Telegraph Company, November 8, 1983.

3. See Luftig and Yelsey, "American Telephone and Telegraph Company." See also Mark D. Luftig and Neil Yelsey, "Bell System—1984 Earnings Update," Salomon Brothers, Inc., Report, December 2, 1983.

4. For a discussion of the four most frequently used cost-of-common-equity methodologies, see Gregory B. Enholm, "A Survey of Four Cost of Equity Methodologies," Madison, Public Service Commission of Wisconsin, September 1983, mimeographed.

5. David S. Evans and Michael Rothschild, "The Impact of Divestiture on the Cost of Capital to the Bell System," in *Breaking Up Bell,* David S. Evans, New York: North-Holland, 1983, 157–89.

6. Professor Charles Linke, University of Illinois, included his survey with his testimony in Illinois and Wisconsin.

7. For a discussion of regulatory strategic planning, see testimony of J. Robert Malko, Wisconsin Power and Light rate case 6680–UR–14, Wisconsin Public Service Commission, June 1984.

5
The Economic and Political Determinants of the Requested-Granted Rate of Return in Public Utility Rate Cases

Homayoun Hajiran
Wheeling College
David R. Kamerschen
John B. Legler
University of Georgia

Several attempts have been made to find the optimal rate of return, but few studies have attempted to analyze how regulatory commissions actually arrive at the allowed rate of return,[1] or to examine the factors affecting the rate of return set by the regulatory commissions in formal regulatory hearings.

The primary objective of this study is to investigate the rate of return a regulated firm will request and the rate it will be granted in a formal hearring. Attempts have been made to provide a better understanding of the regulator, the regulated firm, and the regulatory process itself. Our model is an extension of earlier studies by Joskow (1972a, 1972b, 1972) using New York as an example. This study is based on the regulation of public utilities by the Georgia Public Service Commission. Although telephone and gas company data are used in this study, primary attention is given the electric utility industry, because (1) more complete and consistent data are available for electric companies; (2) it is the utility most commonly considered in previous studies; (3) it is the only type of utility used for estimating one of the three equations in all previous modeling efforts reviewed.

Paul Joskow (1972a, 1972b) conducted pioneering research concerning requested and granted rate of return in a formal regulatory hearing. Using New York state data, Joskow estimated a two-equation behavioral model

---

This chapter is based in part on Homayoun Hajiran, "The Determination of the Requested-Granted Rate of Return: A Case Study in Georgia," unpublished Ph.D. dissertation, University of Georgia, 1982.

of the regulatory process. His data base consisted of twenty rate cases involving major utility companies for the period 1960–70. Joskow's two-equation behavioral model consisted of a commission decision equation and a firm request decision equation. The granted rate of return of the commission decision equation depended on the size and relative reasonableness of the firm's request, the presence or absence of intervenor cost of capital testimony, the type of firm, and its efficiency. The requested rate of return of the firm request equation depended on the embedded cost of debt and the firm's estimate of the equity capital.

In another study, Joskow (1972a, 1973) examined the pricing decisions of regulated firms. This study focused on the decision to request a rate increase or to volunteer a rate decrease. He showed that the decision to file for an increase depended on the growth rate of earnings per share in the current and previous year, on the level of interest coverage ratio in the current year, and on a variable that measures prior expectations of success in the hearing room.

Joskow's results indicated that all coefficients had the expected sign and that all, except the coefficient of the efficiency variable (subjective judgment of the commission concerning the efficiency of the firm), were significant at the 5 percent level. These conclusions will be further discussed in the next section of this chapter.

Roberts, Maddala, and Enholm (1978) performed another requested-granted rate-of-return study. Using Florida data, they extended Joskow's approach by employing alternative econometric analyses—namely, probit and tobit two-stage least squares. They used data on four electric utilities in Florida over the period 1960–76. Although their estimation results were significantly different from Joskow's, only a few coefficients had the expected sign and were significant at the 5 percent level. These results will also be discussed in the next section.

The model used here will be discussed in the third section, it will also include a pricing decision equation for regulated firms, a firm's decision equation, and a commission decision equation. The coefficients of the model were estimated using data for Georgia. Before introducing the econometric model, we will review the institutional, legal, and regulatory setting in Georgia.

The primary regulatory agency in Georgia is the Georgia Public Service Commission (GPSC), created in 1911.[2] Five commissioners are elected by statewide general election to serve six-year terms of office. The powers and duties of the GPSC include setting policy, adapting and enforcing commission rules and regulations for covered public utilities and transportation industries, hearing appeals regarding these rules and regulations, and granting or denying changes in rates and quality of services.

During the period of this study, seventy-seven public utilities were under the jurisdiction of the GPSC, as follows:

Electric	2
Natural gas	4
Telephone	43
Radio and common carrier	26
Telegraph	2

The GPSC is obligated "to assure that all service, utility, and transportation industries under its jurisdiction provide fair, just and reasonable and nondiscriminatory interstate rates and charges. Moreover, any changes in the rates, types, or quality of service being performed must first be approved by the [GPSC] for agencies under its jurisdiction."[3]

Previous Research

The first major studies related to the determination of the rate of return in formal regulatory hearings are in the political science rather than the economics literature. Fred Kort's (1957) application of quantitative methods to the prediction of human events was one of the early studies in this area. His study indicated that it is possible to take some already-decided cases, identify factual elements that influenced the decisions, empirically estimate the numerical values for these elements, and then predict the decisions of the remaining cases in the area specified.

Davis, Dempster, and Wildavsky (1966) introduced the study of the behavior of decision makers in the budgetary process. Their work demonstrated that the budgetary process can be represented by simple models that are stable over periods of time, linear, and stochastic. Their study was concerned with the requests presented in the President's budget for an individual agency and the behavior of Congress with regard to the agency's appropriation. The equations considered for the agency-budget decision were (1) agency requests as a function of the previous year's appropriation; (2) agency requests as a function of the differences between the agency request and the appropriation in the previous year; and (3) agency requests as a function of the previous year's request. Similarly, the equations considered for congressional decision were (1) congressional appropriation as a function of the agency's request to Congress; (2) appropriation as a function of the agency's request and as a function of the deviation from the usual relationship between Congress and the agency in the previous year; and (3) appropriation as a function of the segment of the agency's request that is not part of its appropriation or request for the previous year.

Joskow (1972a, 1972b) made the first attempt to specify and to estimate a simple model of the rate of return. Using Davis, Dempster, and Wildavsky's approach (formulating a model that explains the decision maker's behavior in a budgetary process), he studied the determination of allowed rates of return in formal regulatory hearings. A principal finding of Joskow (1972b, p. 632) was: "Although well defined legal rules for the calculation of the allowed rate of return have not evolved in most regulatory jurisdictions, the results indicate that the regulatory agency, in a consistent fashion, makes use of the information provided to it in the regulatory hearing."

Joskow used a two-equation model of the rate-of-return phase of the regulatory hearing process: (1) a commission decision equation, with granted rate of return (GR) as the dependent variable and (2) a firm request (FR) equation with the requested rate of return as the dependent variable. The model was then stated as follows:

$$\mathrm{GR} = A \cdot X_1 + U_1$$
$$\mathrm{FR} = B \cdot X_2 + U_2$$

where GR = granted rate of return by the commission
FR = requested rate of return by the firm
X_1, X_2 = vectors of explanatory variables
U_1, U_2 = error terms
A, B = vectors of coefficients.

The explanatory variables of the commission decision equation were (1) the firm's requested rate of return, (2) a 0-1 dummy variable for firm testimony on the cost of equity, (3) a variable for the rate of return suggested by an intervenor, (4) a 0-1 dummy variable for whether the firm had been commended for service by the commission, (5) a dummy variable for the type of firm, and (6) a variable measuring the aggressiveness of the firm. Joskow viewed the FR equation as tracing out the average relationship between capital costs, company characteristics, and firm requests. He stated: "Firms that stray above this average may be characterized as being relatively aggressive, trying to get more than the 'typical' firm would request *ceteris paribus*" (Joskow, 1972b, p. 635).

The explanatory variables for the firm's request equation were (1) the embedded cost of debt (2), a dummy variable for the firm's type of capital ownership, and (3) a dummy variable for the type of firm.

Data from twenty rate cases involving major gas or electric companies decided by the New York Public Service Commission during the period 1960–

70 were used in the analysis. In summary, the parameters in the estimated models were

$$GR = -1.14 + 1.10FR + 0.40TE - 0.40IT - 0.20GE - 0.64RE + 0.11EF \quad (5.1)$$

where GR = the estimated granted rate of return
TE = dummy variable representing presence of cost-of-equity testimony
IT = dummy variable representing the presence or absence of an intervenor
GE = dummy variable (gas = 1, electric = 0)
RE = positive residual from the request equation
EF = dummy variable (efficient firm = 1, otherwise = 0).[4]

and

$$FR = 4.32 + 0.64IC = 0.44EG - 0.33OC \quad (5.2)$$

where FR = the estimated requested rate of return
IC = embedded cost of debt (%)
EG = dummy variable (gas = 1, otherwise = 0)
OC = dummy variable (firms with outside capital structure = 1, otherwise = 0).[5]

Joskow found that the requested and the granted rate of return are positively related; the presence of cost of equity testimony will increase, whereas the presence of an intervenor will decrease the allowed rate of return; and higher embedded cost of capital results in a higher requested rate of return by the firm.

Joskow also concluded that an increase of one percentage point in the embedded cost of debt will lead to an increase of 0.64 percentage points in the firm's request (ceteris paribus). The presence of cost-of-capital testimony is worth 0.40 percentage points in the allowed rate of return; the presence of an intervenor will cause a reduction of zero to 0.40 percentage points, depending on the degree of the conflict between the presented testimonies; and it appears that the commission does discount the requests of gas companies more than it does those of electric companies.

Later, Joskow (1973) examined the firm's decision about whether or not to request a rate increase, compared to the question of how much of a rate increase in his previous study. This was an extension of the firm's request equation. He believed that there were two sets of explanatory variables affecting two different decisions about (1) whether or not to apply for an increase, and (2) how much of a rate increase to ask. Joskow's behavioral model (using rule-of-thumb decision making) appears to describe the short-run pricing decision process.

According to Joskow (1973, p. 121), the decision of whether or not to file for an increase is a function of the deviation of actual performance from target performance and an assessment of the chance of getting favorable action from the regulatory authority. Certain financial variables were used for comparing the actual values with the threshold values. Nevertheless, there is a so-called timing problem, and as Joskow (1973, p. 129) pointed out: "There is no way of telling exactly when a firm made its decision to petition for an increase or decrease in rates or exactly what financial information was available to it when it made its decision."

The request decision equation was specified in the form of probit model. The financial variables were as follows:

$$Q = P\,(I_N = 1) = P\,(I_j > I_t)\; I_t \sim N(0,1) \qquad (5.3)$$
$$I_j = 14.47 - 0.14X_1 - 0.22X_2 - 0.71X_3 - 10.62X_4$$

where I_j = threshold values of the decision model
Q = probability that a rate increase will be filed
I_j = decision index
X_1 = growth rate of earnings per share
X_2 = growth rate of earnings per share lagged one period
X_3 = level of interest coverage ratio
X_4 = chance of success [earned rate of return (EROR)/ last allowed rate of return (LROR)].

The estimated coefficients based on sixty-three observations of electric utilities in New York were significant at the .05 level (except for the coverage ratio coefficient). Moreover, the coefficients of the growth rate of earnings per share, earnings per share lagged one period, and the coverage ratio were all negative. This indicates that a deteriorating financial condition leads to a decision to request a rate increase. The coefficient of the chance of success variable was also negative, indicating that a higher ratio of earned to allowed return (X_4) discourages filing a rate increase request. The rate increase and the rate decrease decisions follow similar mathematical and analytical reasoning.

Roberts, Maddala, and Enholm (1978; hereafter referred to as Roberts, 1978) also studied the questions discussed by Joskow concerning what rate of return a regulated firm will request and what rate of return it will be granted in a formal regulatory hearing. They extended Joskow's approach by considering both selectivity and simultaneity issues (these issues are discussed later in this chapter). Using Florida data over the period 1960–76, a total of fifty-nine observations, they estimated parameters using probit and tobit two-stage least squares models.

When they applied Joskow's original request decision model, their results showed that both lagged growth in earnings per share and the ratio of current earnings to the last allowed rate of return had the wrong signs, the current growth in earnings was not significantly different from zero, and only the interest coverage ratio was significant with the correct sign. Therefore, with their methods of estimation (Roberts, 1978, p. 617), all the variables had the correct signs and lower standard errors.

In the Roberts (1978) study, a limitation of Joskow's basic model was pointed out as follows:

> The bargaining process or strategies employed in the request-grant process are entirely captured in the coefficients of the embedded cost of capital and the response of the regulatory commission to the request. The model, therefore, makes no allowance for different strategies to be employed by more (or less) profitable firms. (p. 613)

In an attempt to overcome this deficiency, they introduced a strategy variable (the difference between the earned rate of return and the last rate of return granted by the commission) into their model. This variable was similar to Joskow's chance-of-success variable (earned ROR/last ROR granted) in the decision-to-file equation. Nevertheless, the coefficient of the strategy variable was not significantly different from zero except when the tobit two-stage least squares was used.

With Joskow's basic firm request model, the chance of success variable (X_4) was insignificant. Moreover, none of the coefficients were significantly different from zero when the probit model was used, and the strategy coefficient had a relatively large standard error.

Finally, with Joskow's original commission decision model, an increase of one percentage point in the cost of capital resulted in an increase of only .35 point in allowed rate of return, which is unrealistically low. Thus, the ordinary two-stage least squares estimates and the probit two-stage least squares still produced very low estimates for the requested rate coefficient. However, the tobit two-stage least squares results had the expected signs and relatively low standard errors.

These models did not address the issue of appointed-elected commissioners or the possible influence of the political factor on the decision to file for a rate increase. The basic Joskow model should not be applied across the board to all states if political factors overpower the influence of the financial variables in an elected-commissioner state. Moreover, the Roberts model did not consider any of the dummy variables in Joskow's original model, so the model cannot take account of a commissions treatment of different utilities in their rate request cases.

Empirically, the foregoing models have not utilized data on all types of utilities. Joskow used electric utility and gas utility data, and Roberts used only electric utility data. Data concerning telephone companies' rate requests were not considered. A more complete analysis of the requested-granted process should include the telephone companies as well as gas and electric utilities.

In spite of differences in the models (for example, the econometric analysis, the public utility used, the state, and the time period), the original Joskow model explains the requested-granted process rather well. In this study the political factor and telephone data are incorporated in the original Joskow model and the Roberts version of Joskow's model.

A Three-Equation Model of the Regulatory Process

The behavioral model of the requested-granted rate of return consists of the decision-to-file equation, the firm request equation, and the commission decision equation. Each of these equations explains the dynamic behavior of interested parties at different phases of the requested-granted process. The first and second equations are concerned with the dynamic behavior of the firm. Although closely related, the decisions are influenced by different factors. Moreover, the third equation reflects a compromise between the three conflicting objectives of the parties—the firm, the intervenors, and the commission.

Regulated firms (like other firms) employ rules of thumb in their decision-making process based on indicators that are readily observable. As Simon (1959) emphasizes, decisions are made in an environment of limited information and bounded rationality.

The Decision-to-File Equation

A public utility cannot change the prices of its services independently. However, the firm can make decisions concerning whether or not to file for a rate increase. The burden of proof in justifying the requested rate increase rests with the company. Financial variables usually are major determinants in the decision-to-file equation.

Joskow's (1972a) model was tested using New York data, and the Roberts (1978) model was tested using Florida data. We have applied Georgia data to Joskow's basic behavioral model of the regulatory process, with some necessary modifications. Unlike Florida or New York, the five Georgia commissioners are not appointed but are elected every even year on a staggered-term basis. Therefore, a new variable—a political variable—is included in the model used for Georgia and should be in elected-commissioner states.

The relationship between regulatory action and political factors has recently been recognized and discussed by many economists.[6] For example, in his study of the pricing decision of regulated firms, Joskow (1973, p. 123) mentions the "external pressure from citizens' groups or politicians."

In deciding to file for a rate increase, regulated firms, it is hypothesized, will not pressure the commissions to make a price increase decision before an election. This supports the idea of a relationship between political factors and the granted rate of return, particularly in elected-commissioner states. In other words, the utilities believe that they would receive better treatment after an election than before it. The hypothesis is that before an election, the commissioners tend to take action more favorable to constituents than to the utility.

Therefore, the decision to file can be viewed as two complementary decisions: (1) the whether-or-not decision and (2) the decision about the actual filing date. The choice of actual filing date depends on the hypothesis that the firm will not pressure the commissioners to decide on a rate increase before an election. For example, in Georgia during the period from August of an odd (off-election) year to May of the next (election) year, filing is unlikely to take place, considering the regulatory lag (formally the suspension period) and the election date in November of even years. In other words, if the firm files a rate increase request during this period, considering the regulatory lag of six months, it will force the commissioners to make a decision just before an election. Apparently, as we shall see later, Georgia utilities are convinced that filing during that period is likely to result in an unfavorable attitude toward the requesting firm.

The second decision, whether or not to file for a rate increase, mainly depends on the overall financial condition of the firm. The growth rate of earnings per share, the growth rate of earnings per share lagged one period, and the level of interest coverage ratio before taxes were chosen as major financial variables influencing this decision. The addition of other financial ratios makes the empirical testing unsatisfactory, since most of the conventionally utilized ratios are highly intercorrelated with these three variables. If a firm is in poor financial condition, the growth rate of earnings per share is likely to show a decrease. This might be a result of seasonal factors, short-term business cycles, or a change in some exogeneous variable, and it may not initiate a decision to file for a rate increase. If the financial condition of a company continues to deteriorate through the next period, there is a reason for concern, as the firm realizes that the problem may not be temporary. At this point, the utility is likely to file for a rate increase. As a result, the growth rate of earnings per share lagged one period should be included as a financial variable besides the coverage ratio, which can also affect the financial health of the utility through the bond rating and the cost of debt.

In deciding whether to file for a rate increase, firms seem to get information about the recent behavior of the commission to assess their chances of success. Chance of success is a nonfinancial variable in the equation that we measure by *C*, with *C* defined as EROR/LROR, where EROR is earned rate of return (percent) and LROR is the most recent granted, allowed, or permitted rate of return (percent). Firms deciding to file for a rate increase apparently assess their chances of success in the rate case by comparing their earned rate of return (EROR) and what recently has been granted (LROR). If a firm's EROR is substantially below its LROR, this is an indication that its rate increase should have a very good chance of getting approved. Thus, it is postulated that the higher the ratio *(C)*, the lower the chance of success and, hence, the least likely it is that a firm will file for a rate increase.

Thus, we write the decision-to-file equation as follows:

$$\text{FILE} = A_0 + A_1 \cdot X_1 + A_2 \cdot X_2 + A_3 \cdot X_3 + A_4 \cdot X_4 + u$$

where FILE = a categorical variable equal to unity for every year filing takes place and zero otherwise

X_1 = growth rate of earnings per share (%)

X_2 = growth rate of earnings per share lagged one period (%)

X_3 = interest coverage ratio

X_4 = chance of success *(C)*, or ratio of current rate of return (EROR) to last allowed rate of return (LROR)—that is, EROR/LROR

u = error term.

The Firm Request Equation

Along with the decision to file for a rate increase and, if so, when to file, utilities have to decide how much of a rate increase to request.

The firm-requested rate of return primarily depends on its cost of capital. But only embedded cost of debt will be included in the equation, since, as Joskow (1972a, p. 135) states, "Data on the firm's overall calculated cost of equity is not available in general, and it is treated as an omitted variable." Also, it is hypothesized that the type of utility influences the requested rate of return. For instance, to determine the requested rate of return after embedded cost of debt has been calculated, gas companies may adjust their requested rate differently from electric or telephone utilities. Thus, we write the firm request equation as follows:

$$\text{FIRE} = H_0 + H_1 \cdot Y_1 + H_2 \cdot Y_2 + v$$

where FIRE = firm's requested rate of return (%)
Y_1 = embedded cost of debt (%)
Y_2 = dummy variable for the type of firm
v = error term.

The Commission Decision Equation

After the requested phase of requested-granted rate of return of a regulatory process has been completed, the commissioners must decide how much, if any, of the requested increase should be granted. In deciding the granted rate of return, commissioners must rely on their judgment of the fair and reasonable rate of return. However, Joskow has shown that the granted rate of return can be influenced by the size of the request, the presence or absence of the cost-of-capital testimony, the presence or absence of intervenors' testimonies, the type of firm, and its aggressiveness.

It is postulated that the presence of intervenors tends to reduce the size of the granted rate of return. The extent of this effect depends on the degree of conflict between the firm's and the intervenor's requests. This can be measured as follows:

$$I = (\text{FIRE} - \text{INTR})/\text{FIRE}$$

where I = a variable measuring intervenor's effect
FIRE = firm's requested rate of return
INTR = intervenor's recommended rate of return.

It is also hypothesized that the commissioners tend to treat some utilities differently from others. For instance, if it is believed that gas companies seek a higher risk premium in their requests, the commissioners may try to consider that in deciding on their allowed rate of return.

It is postulated that the commissions discount the request of aggressive firms more than those of others. Further, it is assumed that the cost of debt, weighted by its proportion of the firm's capital structure, is a better index compared to the embedded cost of debt (CD) in measuring the aggressiveness variable. Moreover, more aggressive firms will ask for a relatively higher return to common equity resulting in a relatively higher weighted cost of common equity. In other words, the more aggressive the firm, the larger will be the difference between the requested rate of return and the weighted cost of debt. Hence:

$$S = \text{FIRE} - (\text{PD} \cdot \text{CD})$$

where S = aggressiveness variable (a positive percentage)
PD = percentage of debt in capital structure
CD = the embedded cost of debt (%).

Thus, we may write

$$\text{COGR} = B_0 + B_1 \cdot Z_1 + B_2 \cdot Z_2 + B_3 \cdot Z_3 + B_4 \cdot Z_4 + e$$

where COGR = commission granted rate of return
Z_1 = firm-requested rate of return (FIRE)
Z_2 = dummy variable for the type of utility
Z_3 = intervenor variable *(I)*
Z_4 = aggressiveness variable *(S)*
e = error term.

In summary, the model and a prior expectation of coefficients can be expressed as follows:

$$\begin{aligned} \text{FILE} &= A_0 + A_1 \cdot X_1 + A_2 \cdot X_2 + A_3 \cdot X_3 + A_4 \cdot X_4 + u \\ \text{FIRE} &= H_0 + H_1 \cdot Y_1 + H_2 \cdot Y_2 + v \\ \text{COGR} &= B_0 + B_1 \cdot Z_1 + B_2 \cdot Z_2 + B_3 \cdot Z_3 + B_4 \cdot Z_4 + e \end{aligned}$$

A_1 (negative): The higher the growth rate of earnings per share, the lower the tendency for filing a rate increase request.

A_2(negative): The higher the growth rate of earnings per share lagged one period, the lower the tendency for filing a rate increase request.

A_3(negative): The lower the interest coverage ratio, the higher the tendency to file for a rate increase.

A_4(negative): The higher the ratio of EROR to LROR, the lower the likelihood of receiving favorable treatment and the lower the probability of filing for a rate increase.

H_1(positive): The higher the embedded cost of debt, the higher the requested rate of return.

H_2(positive): If the utility filing is a gas company, the requested rate of return will be higher, as a risk premium is added.

B_1(positive): The commission's granted rate of return will be higher, the higher the amount requested by the firm.

B_2(negative): The commission's granted rate of return will be lower to offset the added risk premium if the requester is a gas utility.

B_3(negative): The commission's granted rate of return will be lower if the requester is an aggressive firm.

B_4(negative): The commission's granted rate of return will be lower if there are intervenors participating.

The Data

The three-equation model was tested using data on five major public utilities in Georgia, including data on two privately owned electric utilities (Georgia Power Company and Savannah Electric and Power Company) and two gas companies (Atlanta Gas Light Company and Gas Light Company of Columbus). The choice of gas and electric companies provides consistency in comparing Georgia results with previous studies. To study the requested-granted behavior of different type public utilities, data on Southern Bell Telephone Company were also tested.

The model equations were estimated using information contained in twenty-seven dockets of major rate cases before the GPSC over the period 1970–81,[7] earnings per share and interest coverage data from *Moody's Public Utility Manual* and *Financial Statistics of Public Utilities,* and the rate of return data from *The Statistics of Class A and B Privately Owned Electric Utilities in the United States* for the period 1968–80.

In estimating the decision-to-file equation, the date the commission will reach a decision is unknown. Therefore, it was assumed that if the rate increase request occurred before the beginning of July, the decision to file would have been based on the company's financial position in the previous year. Similarly, in estimating the commission decision equation, if there was testimony from more than one intervenor, the average recommendations were used. For instance, if the commission staff recommended a return on equity of 12 percent and the consumer utility counsel, a consumer advocate, recommended 11 percent, the average of 11.5 percent was used as the intervenor's recommendation. The differences were small (usually one or two percentage points) and do not appear to pose a problem.

Estimation Techniques

The following equation represents the general form of the firm's and commission's equations:

$$Y = B \cdot X + e$$

where Y = vector of random variables (dependent variables)
X = design matrix (independent variables)
B = vector of coefficients
e = error term.

Ordinary least squares (OLS) was used to estimate the coefficients of the firm's decision and commission's decision equations (see, for example, Schmidt, 1976). In the case of decision to file, the application of the linear regression model is more complex, since the dependent variable may only be two values, zero and unity. A suitable model is one that makes possible the prediction of decision to file (in the form of conditional probability), given the values of independent variables. One alternative is to use the linear probability function by taking the expected values of dependent variables. When dependent variables may be only two values (0, 1), OLS estimates are inefficient, since the two assumptions in the Gauss-Markov theorem will be violated. Furthermore, the probabilistic nature of the decision process makes the assumption of linearity also inappropriate. In the linear probability model, the estimated dependent variables are not restricted between zero and unity. This may result in a problem of predicting dependent variables outside the (0, 1) interval.

In view of the foregoing problems, a probit model was employed to find the estimated coefficients of the decision-to-file equation. Basically, a probit model (Finney, 1964) is a monotonic transformation of the original equation such that the estimated values of the dependent variables are constrained between zero and unity. A normal cumulative probability function can be used for transformation purposes.

Difficulty in estimating the unknown coefficients in the decision-to-file model may occur because information on some variables (firm-requested rate, intervenor's recommendation, and so on.) is not available for those utilities that did not request a rate increase in certain years. This is the so-called selectivity problem. The company has a choice of whether or not to request a rate increase, which in turn determines whether or not it will be included in the subsample of utilities used to estimate different equations in the model.

As Roberts (1978) stated:

> The decision about whether or not to request a rate increase, and, if so, how much of a rate increase to request are interrelated. A firm would decide to ask for a rate increase and incur the costs of preparing the case before the commission only if it were profitable to do so. . . . One observes a firm's request for a rate increase if and only if the firm decides to ask for an increase. (pp. 612–13)

As a result, depending on the degree of sample selectivity bias, the OLS procedure can be biased and inconsistent even if the functional form is linear (see, for example, Heckman, 1976; Wales and Woodland, 1980).

On the other hand, when a firm requests a rate increase, it is an indication that the firm is not satisfied with its current (or expected future) earned rate of return. Thus, the requested rate will be higher than the existing earned rate of return. This requested rate is most likely to be greater than the previous rate. Hence, when a firm did not request an increase, one may assume that it was satisfied with its earned rate of return. In other words, the earned, the requested, and the granted rate of return were implicitly equal. Therefore, the selectivity issue might not be as important as Roberts suspected it would be.[8]

The simultaneity issue arises as the error terms in the firm and the commission decision equations are correlated. However, an appropriate econometric model can correct the problem. In the Roberts model, it was suggested that the selectivity and the simultaneity problems are significant in the requested-granted rate-of-return models. As a result, probit two-stage least squares and tobit two-stage least squares models were used in that study.

Thus, in spite of the foregoing considerations, for simplicity and for the possibility of comparison with previous studies, a linear functional form, a probit model, and ordinary least squares were used in estimating the present model. The results of this and some previous studies should be interpreted with some reservation, depending on the degree to which the following assumptions hold: relatively small selectivity bias, insignificance of the simultaneity problem, no prior information concerning the coefficients, and all of the necessary conditions in the Gauss-Markov theorem, particularly the normality assumption of the error terms distribution.[9]

Estimation Results

The exact Joskow model could not be used in this study, since there was not sufficient information about the commission's opinion of the utility's efficiency and since cost-of-equity testimony was presented in almost all cases. Tables 5–1 and 5–2 contain the maximum likelihood estimates of the probit model for the firm's decision-to-file equation. Column (a) of table 5–1 provides the estimation results using only electric utility data in Georgia. As indicated, all of the coefficients have the correct signs except the coverage ratio. However, none of the coefficients is significantly different from zero.

The comparison among columns (a), (b), and (c) of table 5–1 reveals that there is only one coefficient in three models for Georgia, New York, and Florida that has the correct sign—namely, the unlagged growth rate of earnings per share coefficient. Moreover, there is no common coefficient in the three models that is significantly different from zero.

Table 5–1
Estimation Results of the Decision-to-File Equation Using Only the Electric Utility Data

	(a)	*(b)*	*(c)*	*(d)*	*(e)*
Model	Joskow	Joskow	Joskow	Roberts	Roberts
Intercept (constant)	0.47 (0.09)	14.47 (2.64)	2.75 (0.98)	2.25 (1.82)	3.30 (2.77)
Earnings per share	−0.05 (−0.03)	−0.14 (−2.33)	−0.16 (−0.09)	−0.94 (−0.83)	−2.28 (−1.72)
Earnings per share lagged	−0.40 (−0.30)	−0.22 (−2.20)	5.53 (2.03)		
Interest coverage ratio	0.26 (0.29)	−0.71 (−1.51)	−1.34 (−2.73)	−0.88 (−1.84)	−1.00 (−2.94)
Chance of success (EROR/LROR)	−1.06 (−0.24)	−10.62 (−2.58)	0.97 (0.30)		
Number of observations	19	63	53	26	53
State[a]	Ga.	N.Y.	Fla.	Ga.	Fla.

Note: t-values are in parentheses.

[a]The data for two electric utilities (1967–79) were used for Georgia, the data for eleven electric utilities (1961–69) were used for New York, and the data for four electric utilities (1960–76) were used for Florida.

Table 5–2
Estimation Results of the Decision-to-File Equation: Some Alternative Adapted Joskow Models for Georgia, 1967–79

	(a)	*(b)*	*(c)*	*(d)*
Intercept (constant)	2.06 (1.64)	0.39 (0.89)	0.33 (0.76)	−0.19 (−1.16)
Growth rate of earnings per share	−0.72 (−0.63)	−0.28 (−0.85)	−0.23 (−0.75)	−0.31 (−0.82)
Growth rate of earnings per share lagged	0.97 (0.89)	−0.20 (−0.70)		−0.17 (−0.60)
Interest coverage ratio	−0.85 (−1.74)	−0.16 (−1.40)	−0.16 (−1.35)	
Number of observations	26	65	65	65
Type of data[a]	E	E·T·G	E·T·G	E·T·G

Note: t-values are in parentheses.

[a]E stands for electric utilities, G for gas companies, and T for the telephone company.

Columns (d) and (e) of table 5–1 contain the Roberts version of Joskow's model. The estimated coefficients in these two columns have the anticipated signs. In addition, the coverage ratio coefficient has the correct sign and has a higher level of significance in the Roberts model [column (d)] than in Joskow's model [column (a)].

The results shown in table 5–2—columns (b), (c), and (d)—incorporate telephone company data in addition to gas and electric companies. This results in alternative models in which all of the coefficients have the correct signs but again are not significantly different from zero. In other words, table 5-2 shows the estimated coefficients of the adapted Joskow model when the data for electric, gas, and/or telephone companies in Georgia were used.

One of the factors contributing to these differences between Georgia results and Florida or New York results is the political factor. As mentioned earlier, unlike New York and Florida, which are appointed-commissioner states, Georgia commissioners are elected.

Since the election of commissioners takes place in November of every even year, utilities avoid filing for a rate increase during the preelection period from August 1 of an off-election year to May 31 of an election year. The examination of the data revealed that only two observations out of twenty-seven rate increase request-filing dates were within the preelection period. These two exceptions were Savannah Electric Company's filing on March 23, 1982, and Gas Light Company of Columbus's filing on August 19, 1977. The relatively small service areas of these companies may explain those exceptions. The number of voters in those service areas do not have a significant influence on the reelections of commissioners, who are elected on a statewide basis. Consequently, those companies might have filed assuming that the commissioners' concern was not so great as it could have been if the two big utilities, Georgia Power or Southern Bell, had filed a rate increase request.

Of twenty-seven requests for rate increases during 1970–81 in Georgia, nine were filed in June; four in July; three in January; two each in October, November, March, and May; none in September or February; and only one rate increase was filed in each of the remaining three months.

Hagerman and Ratchford (1978), in their study of the economic and political variables affecting the rate of return on equity, concluded:

> The most interesting result is that whether or not commissioners are elected does not affect the allowed return on equity. This is inconsistent with the conventional wisdom that elected commissioners are responsive to the public interest, i.e., give lower returns. Not only is the dummy variable insignificant, but it even has the wrong sign. (p. 54)

The finding of our study not only supports the so-called conventional wisdom but also may explain why Hagerman and Ratchford reached these conclusions, since their results and conventional wisdom are not necessarily mutually exclusive.

We conclude that the public utilities' decision to file for a rate increase in an elected-commissioner state such as Georgia is primarily influenced by the political variable. Furthermore, utilities avoid putting the commissioners in the position of making a decision on a rate increase case before an election. The point is that because utilities recognize the importance of the political variable and offset its effects (by filing outside the preelection period), there is rather low public pressure on elected commissioners. This low pressure is comparable to that of appointed-commissioner states. Therefore, the commissioners are not compelled to make a lower-return decision, and they behave as in an appointed-commissioner state. This reasoning may explain why the coefficient of the dummy variable for elected-commissioner versus appointed-commissioner states in the Hagerman-Ratchford model had the wrong sign and was not significant. They were trying to capture the difference between two commission forms (appointed and elected), which our study indicates are subjected to similar public pressure and consequently behave the same.

It should be noted that the commissioners' concern about the political variable does not imply that they are overlooking the just and reasonable criterion (as in the *Federal Power Comm.* v. *Hope* case) in their decisions. Rather, as Hilton (1972) states:

> Members of regulatory bodies are typically professional politicians who serve a limited term, usually well under a decade, before going on to the practice of law, a position in private industry, or further elective or appointive offices. . . . Regulators are not automatons, but men and women who go to baseball games, ...take their cats to the veterinarian, and otherwise behave like the rest of us. To put it into economic jargon, they have utility functions like all men, in which they seek to optimize with calculations concerning the present versus the future in light of the costs and rewards presented to them. (pp. 47–48)

Table 5–3 contains the estimation results of the firm's request equation. The results indicate that the coefficients for the embedded cost of debt and for the dummy variable (gas companies = 1) both have their expected signs (positive) and are significant at the .01 and .05 levels, respectively. But the coefficient for the outside-capital-structure dummy variable had the wrong sign, and it was not significantly different from zero. Joskow assumed that companies owned by a corporation outside the state tend to discount their requests.

Furthermore, the values of the coefficients suggest that gas and electric utilities in Georgia request .94 percent of their embedded cost of debt plus

Table 5–3
Estimation Results of the Firm Request Equation

	(a)	*(b)*	*(c)*	*(d)*
Model	Joskow	Joskow	Roberts	Roberts
Intercept (constant)	2.32 (2.03)	4.32 (7.78)	3.40 (2.50)	3.51 (1.29)
Embedded cost of debt	0.94 (7.33)	0.64 (5.17)	0.82 (5.00)	0.87 (2.63)
Dummy variable (gas = 1)	0.78 (2.22)	0.44 (2.33)		
Dummy variable (outside capital structure = 1)	0.28 (0.78)	−0.33 (−1.64)		
LROR minus EROR			0.04 (0.25)	−0.26 (−0.20)
R^2	0.806	0.647	0.799	0.438
Number of observations	19	20	10	13
Type of data[a]	E · G	E · G	E	E
State[b]	Ga.	N.Y.	Ga.	Fla.

Note: t-values are in parentheses.

[a]E stands for electric utilities and G for gas companies.

[b]Georgia data covers the period 1970–81, New York data 1960–70, and Florida data 1960–76.

2.32 points (intercept). But gas companies adjust this total by adding an additional .78 points (risk premium). There is also another adjustment of .28 points for firms with out-of-state ownership (outside capital structure). The estimation results for Georgia data using the Roberts model is shown in column (c) of table 5–3. Nevertheless, the Roberts model did not include any dummy variables and employed only electric utility data. The results show that the coefficient for the embedded cost of capital variable has the correct sign and is significant at the .01 level.

An interesting observation is suggested in table 5–4, columns (b) and (c). Compared to gas or electric companies, the results indicate that the telephone company adjusts its request higher than other utilities. In column (c), electric, gas, and telephone data were used. The coefficients suggest that all utilities request 1.01 percent of their embedded cost of debt plus 1.98 points (intercept). Gas companies adjust this total by requesting an additional .61 points, whereas the telephone company adjusts the former total by an additional .92 points, or a total of 1.73 points above electric utilities.

Joskow referred to the estimated coefficient for gas companies' dummy variable of .44 [column (b) of table 5–3] as a risk premium. However, in our study, the adjustment by the telephone company is not a risk-adjustment

Table 5–4
Estimation Results of the Firm Request Equation: Some Alternative Adapted Joskow Models for Georgia, 1970–81

	(a)	*(b)*	*(c)*
Intercept (constant)	2.78 (2.87)	0.59 (0.33)	1.98 (1.98)
Embedded cost of debt	0.91 (7.75)	1.26 (5.70)	1.01 (8.27)
Dummy variable (gas = 1)	0.61 (2.33)		0.61 (2.04)
Dummy variable (telephone = 1)		1.31 (3.03)	1.73 (5.22)
R^2	0.799	0.769	0.789
Number of observations	19	13	25
Type of data[a]	E·G	G·T	E·T·G

Note: t-values are in parentheses.
[a]E stands for electric utilities, G for gas companies, and T for telephone companies.

premium. The reason lies in the way data were grouped. Considering the capital structure and the cost of capital, gas and electric companies are more homogeneous than telephone and electric or telephone and gas companies.

Table 5–5 indicates the capital structure and the cost of each component for gas, electric, and telephone companies as presented in twenty-seven rate cases during 1970–81 in Georgia. Column (a) indicates that the proportion of debt in the capital structure for gas and electric utilities has been in the 50 to 60 percent range, compared to 40 to 49 percent for the telephone company during 1970–81. Therefore, the financial risk of the telephone company has been lower than that for the gas or electric companies. The lower cost of debt for the telephone company, compared to electric utilities, is consistent with this concept (AAA bond rating for Southern Bell, BBB for Georgia Power, and BB for Savannah Electric Company).

Consequently, since the requested rate of return is stated as a function of embedded cost of debt, the estimation results, when all utilities' data are used, will show that the telephone company has included a compensating premium to its requested rate above gas and electric utilities. Therefore, the telephone company adds this compensating premium not for being riskier but rather for being less risky than other utilities. The Joskow and Roberts models did not address this issue, since Joskow used electric and gas company data and Roberts used only electric utility data.

Table 5–6 contains the estimation results of the coefficients in the commission decision equation. Column (a) includes the results of Joskow's model applied to Georgia data. All of the coefficients have the expected signs except

Table 5–5
The Capital Structure of Public Utilities in Georgia
(percentages)

		Debt		*Preferred Stock*		*Common Equity*	
Company	*Year*	*(a) Portion*	*(b) Cost*	*(c) Portion*	*(d) Cost*	*(e) Portion*	*(f) Granted*
Georgia Power	81	60.41	9.32	10.67	9.39	28.92	15.50
	78	57.85	8.22	11.00	8.42	31.17	13.50
	77	55.38	7.98	12.90	8.26	31.72	12.25
	74	59.00	7.68	10.00	7.39	31.00	12.75
	73	52.10	6.96	10.42	6.65	31.68	12.00
	72	56.50	6.74	11.40	6.69	32.10	11.50
	71	57.10	6.27	10.70	6.13	32.20	11.50
Savannah Electric	80	73.40	9.70	8.28	10.81	28.32	14.40
	79	65.90	9.14	8.90	10.78	25.20	12.75
	77	61.99	9.19	11.46	10.71	26.55	12.75
	75	51.84	8.80	9.28	9.12	27.73	11.57
	72	47.48	6.30	7.00	6.95	32.13	12.00
Atlanta Gas	81	51.33	9.11	10.10	7.60	38.37	16.00
	78	54.37	8.94	10.54	7.57	35.09	14.00
	76	51.44	7.17	13.80	7.52	34.76	13.25
	72	54.10	6.08	9.50	6.24	36.40	12.00
Columbus Gas	79	41.38	8.66	8.00	6.00	44.70	14.25
	77	52.73	8.43	7.54	6.00	39.73	13.75
	74	61.45	7.75	7.25	6.00	31.30	13.00
Southern Bell	81	43.56	8.39	2.12	7.79	54.32	14.25
	80	44.28	7.83	2.29	7.79	53.43	13.50
	79	46.19	7.32	2.56	7.70	51.25	12.00
	76	48.70	6.90	4.00	7.80	47.30	12.50
	74	49.10	7.06	5.00	7.94	45.90	11.50
	71	46.76	6.04	3.74	8.01	49.48	10.90

Note: Where the total does not add to 100 percent, the difference is attributable to short-term debt.

the dummy variable for gas companies. However, coefficients of the gas dummy variable and the aggressiveness variable are not significantly different from zero.

The coefficients suggest that the commission granted 86 percent of the requested rate plus 1.26 points (intercept) with the following considerations:

1. Upward adjustment (.03) for gas companies;
2. Discounting (.12) the requests of the aggressive firms (not significant);
3. Discounting (6.05) the requests based on the intervenor's recommendation.

The estimation results of the Roberts model [column (c) in table 5–6] show that the firm-requested-rate coefficient has the correct sign (positive)

Table 5–6
Estimation Results of the Commission Decision Equation

	(a)	*(b)*	*(c)*	*(d)*
Model	Joskow	Joskow	Roberts	Roberts
Intercept (constant)	1.26 (2.15)	−1.14 (−1.70)	2.26 (2.03)	3.45 (2.65)
Firm's rate of return request	0.86 (15.60)	1.10 (10.85)	0.72 (6.55)	0.54 (3.83)
Dummy variable (gas = 1)	0.03 (0.22)	−0.20 (−2.18)		
Aggressiveness variable, positive residuals	−0.12 (−0.55)	−0.64 (−2.39)		
Intervenor's variable (1 − INTR/FIRE)	−6.05 (−2.93)	−0.40 (−3.89)		
EROR minus LROR			0.03 (0.35)	−0.07 (−0.93)
R^2	0.968	0.945	0.874	—
Number of observations	16	20	10	13
Type of data[a]	E·G	E·G	E	E
State[b]	Ga.	N.Y.	Ga.	Fla.

Note: t-values are in parentheses.

[a]E stands for electric utilities and G for gas companies.

[b]Two coefficients are not included in the Georgia model but were included in the New York model: (1) the dummy variable coefficient for cost-of-capital testimony (0.40) and (2) the dummy variable coefficient for the efficiency variable (0.11).

but that the strategy variable (EROR minus LROR) coefficient does not have the correct sign. Furthermore, the former is not significantly different from zero.

In table 5–7, columns (a), (b), (c), and (d) are some alternative adapted Joskow models in applied to data from different groups of utilities in Georgia. For instance, in column (d), gas, electric, and telephone company data were used for estimating the coefficients. Again, all of the coefficients have the correct signs and are significantly different from zero at .05 level except the dummy variables (the firm request coefficient was significant at the .01 level). Also, it seems that the commission discounted the requests of risky firms (to offset their added risk premium). Furthermore, electric companies' requested rate of return was discounted more than that of gas companies. However, the standard errors of the foregoing dummy variables are relatively large.

It also appears that the commission discounted the aggressive firm requests consistently. Moreover, intervenors played an effective role in rate cases in Georgia. For instance, if the requested rate of return was 12 percent

Table 5–7
Estimation Results of the Commission Decision Equation: Some Alternative Adapted Joskow Models for Georgia, 1970–81

	(a)	*(b)*	*(c)*	*(d)*
Intercept (constant)	1.47 (2.98)	0.92 (2.68)	1.96 (3.09)	1.79 (3.72)
Firm's rate-of-return request	0.91 (9.35)	1.20 (13.50)	0.94 (10.11)	0.97 (12.05)
Dummy variable (gas = 1)	0.15 (0.77)	−0.15 (−1.48)		−0.04 (−0.24)
Dummy variable (electric = 1)			−0.29 (−1.00)	−0.29 (−1.22)
Aggressiveness variable (FIRE − PD·CD)	−0.15 (−0.78)	−0.55 (−5.50)	−0.27 (−1.91)	−0.27 (−2.28)
Intervenor's variable (1 − INTR/FIRE)	−5.62 (−2.60)	−0.63 (−0.60)	−3.40 (−1.93)	−3.59 (−2.40)
R^2	0.969	0.996	0.962	0.970
Number of observations	16	10	16	21
Type of data[a]	E·G	G·T	E·T	E·T·G

Note: t-values are in parentheses.
[a]E stands for electric utilities, G for gas companies, and T for telephone companies.

and the intervenors recommended 10 percent, the effect of this discrepancy resulted in a reduction of .937 points [using column (a) in table 5–7] from the granted rate of return; that is, (1-10/12) − 5.62 = .937.

Table 5–8 contains the estimation results for the commission decision model when the intercept was omitted from the equation. The coefficients suggest that in the case of gas companies, 28 percent of the firm's request plus 71 percent of the intervenor's requests composed the granted rate of return; in the case of electric companies, 29 percent of the firm's request plus 69 percent of the intervenor's request composed the granted rate of return. However, the telephone company's granted rate of return depended mainly on how much it requested (94 percent of its request) rather than the intervenor's request. In other words, it seems that the variation in granted rate is explained mainly by the variation in intervenor's request for electric and gas companies and by the variation in firm's request for the telephone company. Using the foregoing coefficients, the result of a one percentage point increase in overall embedded cost of debt (ceteris paribus)[10] can be traced in the model as follows:

1. Column (a) of table 5–4 shows that electric companies will increase their requested rate of return by .91 points.

Table 5–8
Estimation Results of the Commission Decision Equation, Some Alternative Adapted Joskow Models for Georgia with No Intercept

	(a)	*(b)*	*(c)*	*(d)*
Firm's rate-of-return request	0.295 (1.41)	0.278 (1.26)	0.943 (2.84)	0.423 (2.59)
Intervenor's recommendation	0.695 (3.11)	0.711 (2.96)	−0.058 (−0.15)	0.557 (3.20)
Dummy variable (Gas = 1)				0.01 (−0.08)
Dummy variable (Telephone = 1)				−0.21 (−1.09)
R^2	0.9998	0.9997	0.9994	0.9993
Number of observations	11	5	5	21
Type of data[a]	E	G	T	E·T·G

E stands for electric, (G) for gas and (T) for telephone companies.
Note: t-values are in parentheses.

2. Column (a) of table 5–4 shows that gas companies will increase their request by .61 points above the electric companies, or a total of 1.52 points.
3. Column (a) of table 5–7 indicates that the granted rate of return to electric companies will increase by .83 points.
4. Column (a) of table 5–7 indicates that the granted rate of return to gas companies will increase by .98 points.

Summary and Conclusions

The research reported here began by examining behavioral models of the requested-granted rate of return in a regulatory process. The pioneering study by Joskow in the state of New York and a later model developed by Roberts, Maddala, and Enholm in the state of Florida were used as the basis for a test of the behavior of parties involved in the regulatory process in the state of Georgia.

A model similar to Joskow's, with some adjustments to make it more applicable to the Georgia data, was developed. Basically, this model consists of three equations:

1. It was postulated that the decision to file for a rate increase depended on the growth of earnings per share, the growth rate of earnings per share

lagged one period, the coverage ratio (financial variables), and the chance of success in the hearing room (a nonfinancial variable).

Estimation results showed that all of the coefficients, though generally insignificant, had the expected signs in most of the alternative models whether the models used electric utilities alone or were based on different combinations of utility data—that is, telephone and gas, electric and gas, and electric, gas, and telephone. However, only the growth rate of earnings per share coefficient consistently had the negative sign in the different models examined.

We hypothesized that there is a political factor that influences the decision about the actual filing date of a rate increase. Unlike New York or Florida, where the model was tested earlier, in Georgia the commissioners are not appointed but are elected in November of every even year. Therefore, during a preelection period from August 1 of an odd year to May 31 of next year, the likelihood of filing a rate request is minute. If the filing takes place during a preelection period, considering the regulatory lag, the commissioners would be forced to make a decision just before an election. Hence, it is likely that the intervenor's recommended rate would receive a more favorable response than the utility's requested rate of return.

The data support the foregoing argument. During 1970–81, only two out of twenty-seven rate increase cases were filed within the preelection period. Moreover, the majority of the rate requests were filed in June and July, the only two months outside the preelection period in any year. Therefore, in Georgia, where commissioners are elected, it seems that the date of a utility's decision to file for a rate increase has been influenced by political factors more than by financial variables.

2. The firm decision about the size of the rate increase was explored in the second equation. The estimation results for Georgia were very similar to those of New York and Florida. The embedded cost of debt coefficient and the gas company's dummy variable (risk premium) had the correct signs (positive) and were significantly different from zero at the .01 and .05 levels, respectively.

3. The granted rate of return in a formal regulatory process is an outcome of a compromise between three involved parties (the firm, the commission, and the intervenor) with rather conflicting and perhaps mutually exclusive objectives. The commission must use its judgment and available information (based on the presented testimonies) to decide on what the fair rate of return must be in each rate case. In arriving at that decision, a primary objective is to protect the public interest.

There is no set methodology used to decide on the merits of a rate increase proposal. Still, the results of New York, Florida, and Georgia studies indicate that the commissions act rather consistently from case to case and from state to state (at least in the states studied). The commission decision equation postulates that the granted rate of return depends on the requested rate of return by the firm, the rate recommended by intervenors, the type of firm requesting the rate increase, and the aggressiveness of the firm's request.

The requested rate of return coefficient had the expected sign (positive) and was significant at the .01 level, as in other studies. The coefficient of the dummy variable for gas companies had the wrong sign and was not significantly different from zero. The coefficient for the firm's aggressiveness variable had the correct sign (negative), but it was not significantly different from zero. The coefficient for the intervenor's variable had the expected sign (negative), and it was significant at the .01 level. The results for Georgia were similar to Joskow's results for New York, with few exceptions. Furthermore, when telephone data were added to the gas and electric data, a closer resemblance was observed.

In spite of rather poor results of the decision-to-file equation—largely, we believe, because New York is an appointed-commissioner state whereas Georgia is an elected-commissioner state—we concluded that Joskow's basic model performed rather well in Georgia.

Notes

1. Unless otherwise specified, the overall rate of return (vis-à-vis the rate of return on equity) is what we shall refer to as the rate of return.

2. This discussion draws heavily on Jackson (1975).

3. See Jackson (1975, p. 90).

4. Joskow's *t*-values are reported in table 5–6.

5. Joskow's *t*-values are reported in table 5–3.

6. See, for example, Hagerman and Ratchford (1978); Peltzman (1971, 1976); Kamerschen, Kau, and Paul (1982).

7. The rate increase cases used in this study consisted of seven cases for Georgia Power Company, five for Savannah Electric, five for Atlanta Gas Light, four for Gas Light Company of Columbus, and six for Southern Bell Company.

8. In some studies in which the selectivity issue was considered (Heckman, 1976, p. 485), results showed that the selectivity bias was not significant.

9. If any of these assumptions are violated, and for a detailed discussion concerning the consequences of the OLS estimation method, see Judge, Griffiths, Hill, and Lee (1980).

10. The assumption of ceteris paribus is rather a strong assumption, since a change in embedded cost of debt will eventually influence other variables in the model—namely, the intervenor's variable and the aggressiveness variable. On the other hand, one can assume that in the example, the intervenor's request and the firm's request will increase proportionally and will leave the intervenor's variable unaffected (1 − INTR/FIRE). Moreover, the contribution of the aggressiveness variable to the model was not significant.

References

Davis, Otto; Dempster, M.; and Wildavsky, Aaron. "A Theory of the Budgetary Process." *American Political Science Review* . 60(September 1966): 122–53.

Federal Power Comm. v. Hope Natural Gas Co., 320 U.S. 591 (1944).

Financial Statistics of Public Utilities. Illinois: Turner and Associates, 1967–81. Electric and Gas Operating Companies. Bloomingdale, Illinois.

Finney, David J. *Probit Analysis,* 2nd ed. Cambridge: Cambridge University Press, 1964.

Georgia Public Service Commission Report. Atlanta: Georgia Public Service Commission, 1967–80.

Hagerman, Robert L., and Ratchford, Brian T. Some Determinations of Allowed Rates of Return on Equity to Electric Utilities." *Bell Journal of Economics,* Vol. 9 (Spring 1978): 46–55.

Hajiran, Homayoun. "The Determination of the Requested-Granted Rate of Return: A Case Study in Georgia." Unpublished Ph.D. dissertation, University of Georgia, 1982.

Hass, Jerome E.; Mitchell, Edward S.; and Stone, Bernell K. *Financing the Energy Industry.* Cambridge, Mass.: Ballinger, 1974.

Heckman, J. "The Common Structure of Statistical Models of Truncation, Sample Selection and Limited Dependent Variables and a Simple Estimator for Such Models." *Annals of Economics and Social Measurement* Vol. 5 (Fall 1976): 475–92.

Hilton, George W. "The Basic Behavior of Regulatory Commissions." *American Economic Review* 62(May 1972): 47–54.

Jackson, Edwin. *Hand Book of Georgia State Agencies.* Athens: University of Georgia Institution of Government, 1975.

Joskow, Paul. "A Behavioral Theory of Public Utility Regulation." Unpublished Ph.D. dissertation, Yale University, 1972a.

———. "The Determination of the Allowed Rate of Return in a Formal Regulatory Hearing." *Bell Journal of Economics* 3(Autumn 1972b): 632–44.

———. "Pricing Decisions of Regulated Firms: A Behavioral Approach." *Bell Journal of Economics* 4(Spring 1973): 118–40.

Judge, George G.; Griffiths, William E.; Hill, R. Carter; and Lee, Tsoung-Chao. *The Theory and Practice of Econometrics.* Wiley Series in Probability and Mathematical Statistics. New York: John Wiley and Sons Inc., 1980.

Kamerschen, David R.; Kau, James B.; and Paul, Chris W., II. "Political Determinations of Rates in Electric Utilities." *Policy Studies Review* 3(February 1982): 546–56.

Kort, Fred. "Predicting Supreme Court Decisions Mathematically: A Quantitative Analysis of the 'Right to Counsel' Cases." *American Political Science Review* 51(March 1957): 1–12.

Moody's Public Utility Manual. vol. 2. New York: Moody's, 1981.

Peltzman, Sam. "Pricing in Public and Private Enterprises: Electric Utilities in the Unites States." *Journal of Law and Economics* (1971) Vol. 14, 109–47.

———. "Toward a More General Theory of Regulation." *Journal of Law and Economics* (August 1976) Vol. 19, 211–40.

Roberts, Blaine; Maddala, G.; and Enholm, Gregory. "Determination of the Requested Rate of Return and the Rate of Return Granted in a Formal Regulatory Process." *Bell Journal of Economics* 9(Autumn 1978): 611–21.

Schmidt, Peter. *Econometrics.* New York: Marcel Dekker, 1976.

Simon, Herbert. "Theories of Decision Making in Economics and Behavioral Science." *American Economic Review* 49(June 1959): 253–83.

Standard and Poor's Industry Surveys vol. 2. New York: Standard and Poor's Corporation (October 1981).

The Statistics of Class A and B Privately Owned Electric Utilities in the United States. FPC Publication, 1970-1979. Washington, DC: U.S. Government Printing Office, 1980.

Wales, T.J., and Woodland, A.D. "Sample Selectivity and the Estimation of Labor Supply Functions." *International Economic Review* 21(June 1980): 437–68.

Part II
The Companies' Perspective

In this section we get three different perspectives on access charges from AT&T, OCCs, and LECs. There is only one common element in their views: each regards the present regulatory process as flawed in general and the prevailing method of calibrating access charges as flawed in particular. However, each of the authors recognizes that enhanced deregulation carries with it both benefits and costs.

Roy A. Billinghurst shows that some of the so-called solutions to access charge problems are equivalent to a tax on knowledge. His particular emphasis is on two alternative taxing proposals involving levies, the first on a basis that is independent of the usage of the exchange carriers network, and the second on the basis of all the market will bear. In both cases, in his judgment, there is a lack of economic pricing for telecommunication services. In particular, he believes that the capacity charges being proposed to solve uneconomic bypass encourage the development of specialized networks and institutionalize the imposition of a "knowledge" tax. For the greatest social value, access prices, as all prices, should be set to equal marginal costs, with second-best adjustments to meet total revenue requirements in accordance with Ramsey's (inverse elasticity) rules.

Lawrence R. Weber discusses how local-exchange companies' (LECs) revenue streams have been affected by restricting the intra-LATA market to LECs and by establishing access charges to produce revenues equal to those previously derived from inter-LATA toll and extending 35 to 55 percent discounts on access charges to OCCs. He submits that through regulatory loopholes in the intra-LATA market, resellers and OCCs are eroding LEC revenues, thereby producing enhanced access charges to authorized intrastate carriers and their customers or some form of higher local rates. Similarly, discounted access charges to OCCs are eroding the revenues of LECs. He, too, advocates cost-dominated pricing to remedy this situation.

James D. Ferguson believes that OCCs have three inherent disadvantages when unequal access prevails: (1) additional costs with Feature Group A connections; (2) inconvenience in making Feature Group A calls; and (3) limits on the quantum of customers the OCCs can serve. He discusses these perceived disadvantages in some detail. Although he believes that access charges are too high, he advocates a differential access charge for OCCs until they become truly equal with AT&T. Thus, he proposes setting access charges on the basis of not only costs but also value of service and technological efficiency.

Alan K. Price provides a perspective on access charges from the point of view of a LEC. He feels that the FCC's initial access charge plan—which was, in his judgment, designed to promote competition, prevent bypass, and ensure the continued availability of affordable local-exchange service—was superior to the resulting compromise plan. For instance, he is critical of the fact that existing tariffs ignore individual market characteristics and require that every use of time be assessed at the same rate, regardless of the volume of calls, time of day, direction of the calls, and so forth. He suggests that the majority of costs associated with the provision of a subscriber's local loop be removed from interstate charges and that local-exchange service rates necessarily be increased.

Frank Skinner discusses what he perceives as the agenda of unfinished business of the telecommunications industry. It includes effective provision of equal access, development of meaningful pricing options, recovery of the fixed costs of network accesses, and depreciation rates for the recovery of capital investment that reflect the competitive market plan. He develops, with considerable detail, logic, and emotion, this unfinished agenda and tells how it can be finished.

Ward H. White discusses the ongoing transition in the telecommunications industry from the perspective of LECs. He makes several suggestions on how the industry can be made better. They include reduced regulations and reporting requirements, increased pricing flexibility, fewer structural and jurisdictional separations, an implemented capital recovery program, and merged federal and state regulatory requirements.

6
Repeal the Knowledge Tax

Roy A. Billinghurst
AT&T Communications

For those of us who have watched our industry wander through that cryptic maze called access charges, there would seem to be few surprises left. However, I am both surprised and alarmed by some of the new hazards that are coming into view.

I am surprised because even some of the Bell Operating Companies are creating these hazards, and I am alarmed because if the hazards remain in place, they will surely jeopardize some of the greatest potential developments of our emerging information society.

I strongly believe that these hazardous policies are a threat to the well-being of our society, since they amount to a tax on the distribution and expansion of knowledge. I hope to convince you that this threat is real and to enlist your support to *repeal the knowledge tax*.

I suspect that some of you already have that uneasy feeling that either you or I must be in the wrong book. What does a curious concept such as a "knowledge tax" have to do with access charges? Well, rest assured, neither of us is in the wrong book. I will explain why some of the proposed "solutions" to the access charge dilemma are tantamount to a tax on knowledge. I will focus on two of the most disturbing developments: (1) interexchange carrier capacity charges and (2) charging all the market will bear for carrier access service.

Carrier capacity charges is a generic term I will use for any access charge method that levies access charges on interexchange carriers in a way that is independent of their usage of the exchange carrier's network. *Charging all the market will bear* is self-evident, but I will expand on it.

If I were to rename this chapter "Stop Uneconomic Bypass," most of you would immediately agree that I had finally stumbled onto the real issue and that anyone can see that this is the heart of the access charge debate. I do not disagree that uneconomic bypass is the heart, the so-called gut issue, of access charges. However, if uneconomic bypass is the heart of the issue, then uneconomic pricing of telecommunications services is its soul! In other words, if telecommunications prices are not eventually aligned with economic costs, the heart of the public switched network in this country may be kept beating through the miracle of modern subsidies (that is, capacity taxes), but its ability to handle the expanding needs of an information society will be lost.

This fundamental issue is usually ignored in the heat of arguments on uneconomic bypass and universal service. Furthermore, voices ranging from BOCs to consumer groups continue to press for the recovery of significant amounts of fixed network costs through an end-user usage rate, although the justifications for such an uneconomic policy range from a pragmatic "charge all the market will bear" attitude to unfounded concerns that low-income and low-use customers will be denied access to the public networks by economic pricing rules.

It is a sad commentary on the understanding of basic economic forces in our country that such diverse groups advocate telecommunications pricing policies that either ignore established economic principles or attempt to show that economic pricing rules are inconsistent with notions of social equity. As a result of this confusion, uneconomic bypass is being improperly positioned by many parties as the underlying culprit.

I do not mean to trivialize or underplay the potential damage of uneconomic bypass, and I am not suggesting that its well-documented evils be ignored or even placed in the background. On the contrary, I hope that they stay in the forefront of the access charge debates until the real cause of uneconomic bypass, the lack of economic pricing for telecommunications services, is rooted out.

Uneconomic bypass is the shrill marketplace signal that telecommunications prices are highly distorted. Unfortunately, there seems to be gathering momentum to artificially mute this distress signal before the problems that it heralds are fixed.

There is a natural tendency for exchange carriers and regulators to focus strictly on uneconomic bypass and to attempt various forms of cosmetic surgery that may have been partially effective in other utilities (energy, transportation, and so on). However, the public communications network is not like other utilities. Although it possesses a financial body just as any other utility does, it is unique in that its body contains the "nervous system" of our modern society.

Most of the easy solutions to uneconomic bypass merely put the financial body on a life-support system and then hope for the best. In the communications industry, the body may be saved, but its ability to function as society's nervous system will surely be impaired. If the disease of uneconomic telecommunications pricing is not fixed, all physical and financial processes of the public network may appear to be fine, but its brain will be dead: the public network will still transmit all the information needed for routine social maintenance, but its higher-order functions—its ability to enhance and expand knowledge and human well-being—will be gone.

Therefore, it is vital to understand that uneconomic bypass is not the only evil associated with current access pricing and that it may not even be the worst evil. If some of the current schemes to stop uneconomic bypass through interexchange carrier capacity charges and other taxing methods are not accompanied by genuine pricing reforms, many of the life-enriching benefits of the information age will be available only to a select few. As a result, the flow and advancement of knowledge in our society will be restricted just as surely as if an explicit tax had been imposed on knowledge itself.

Virtually every study of telecommunications industry pricing shows that the current system of recovering non–traffic-sensitive costs through usage rates causes enormous losses in economic efficiency. Put more simply, artificially high long-distance rates caused by carrier access charges substantially reduce the information transactions over the public network. In the past, this repressed information flow primarily caused a diminution of routine social contacts. Although this was unfortunate and sometimes painful, it was difficult to view it as a major catastrophe.

But in the future, the combination of intelligent machines and low-cost intercity communications promises an enormous advancement in the social value of information transmitted over the public networks. For instance, we are just beginning to see the social potential for on-line interaction with vast repositories of knowledge, various types of expert systems, computer-based educational systems, and real-time interactions between common interest groups in widely dispersed sections of the country.

Unfortunately, the only way that such knowledge exchanges can be made economically is to use the public data networks, such as TYMNET or TELENET. If the public voice networks are used, the customer charges are several times greater. Although part of this difference is the greater technical efficiency of the so-called packet-switched networks, the biggest difference is the non–traffic-sensitive cost subsidy tax on information sent over the public voice networks.

For instance, the network transport cost of exchanging knowledge with intelligent machines is currently as low as $2 per hour. This can be compared with the costs of using the intelligent machines themselves, which is also as low as $2 per hour. Finally, compare either of these costs with the typical non–traffic-sensitive cost loading on the public voice network, which is at least $6 an hour; in some states, it is over $10 an hour.

I suspect that one of the reasons that the public data networks developed as they did is because of the gross uneconomic pricing of information exchanged over the public voice networks. Of course, if it were not for that development, the debilitating effects of the knowledge tax would have been much more obvious than they are today. But unless this tax is quickly eliminated, the enormous potential for the public networks to bring the benefits of a knowledge network to all people will be lost.

In the place of integrated voice and data knowledge networks will be sophisticated data networks serving major population centers, and a ubiquitous, less sophisticated, and very expensive public (voice) network will serve everyone else. The people and the services that can use the sophisticated networks to plug into the information age will be sensibly avoiding the knowledge tax. However, the overall cost to society of inefficient and possible redundant networks will be substantial. More important, the people and services that cannot use the sophisticated networks because of their fairly narrow scope and serving area will be subject to the full force of the knowledge tax.

It is fairly clear that people who already recognize the life-enriching services of the information age will continue to use them, although at a much lower level than they would if the services were economically priced. Unfortunately, the combination of specialized but nonubiquitous data networks and the inflated information transfer prices in the traditional voice networks will frustrate the development of the large markets needed to drive down the unit costs of knowledge-processing devices (for instance, centralized interactive data bases). The lost opportunity to make knowledge-enhancement systems affordable and available to everyone may drive an information wedge into our society. This wedge will further separate the information-poor from the information-rich and will reduce the well-being of both groups.

The capacity charges being proposed to solve the problems of uneconomic bypass are especially disturbing, since they will encourage the development of specialized networks and institutionalize the imposition of a knowledge tax. But if policymakers have the foresight to reject them or at least combine them with clear-cut access pricing reforms, they can repeal the knowlege tax at the same time that they effectively stop uneconomic bypass. These reforms must be extensive and may cause inconvenience for the most highly subsidized customer groups, but no one is suggesting that economic pricing be implemented overnight or that it be implemented in a way that jeopardizes universal service. Methods for addressing both of these concerns are well known and available to all regulators.

This country is currently suffering through a difficult industrial and societal change to ensure that the most powerful economic force available, the competitive market, can produce innovative and low-cost intercity information transport services. It will be a cruel hoax on society to encumber the largest single input cost (that is, access charges) of intercity carriers with markups or taxes that could not possibly exist in a competitive market. It does not matter whether these taxes are called NTS cost support, fair compensation for the use of valuable resources, or contribution to cover the fixed overhead costs of exchange carriers. The evidence is overwhelming that they lead to economic inefficiencies, uneconomic bypass, and a virtual tax on the enhancement of knowledge in our society.

Does this mean that carrier access charges should get a free ride and make no contribution to the common overhead costs of exchange carriers? No, it does not! However, it also does not mean that exchange carriers should charge all the market will bear for carrier access service. There are well-known economic methods for determining the common costs of exchange carriers, and these methods clearly show that most non–traffic-sensitive costs are neither common nor joint costs. There are also well-known methods for developing service contribution levels that will do minimum damage to economic efficiency and the flow of information in our society.

The notion of charging all the market will bear is often called market-based pricing by exchange carriers. On occasion, AT&T also uses this phrase to describe the proper way to price carrier access service. However, when AT&T uses the phrase, we are talking about the price levels for carrier access service that would occur if the access market were actively competitive. That is, we are talking about basing carrier access rates on the marginal cost of usage. The exchange carriers use this term to mean something quite different—namely, charging all the market will bear.

To maximize economic efficiency and minimize uneconomic bypass, exchange carriers should set carrier access prices (and all other prices) to equal marginal costs. Of course, some adjustments would be needed to meet total revenue requirements, and these should be made in accordance with the Ramsey inverse elasticity rule.

However, exchange carriers are not advocating this process when they recommend market-based pricing. They want to set the price for carrier access at the point at which their marginal revenue from carrier access service equals their marginal cost. If the access market were actively competitive, this procedure would set the price equal to the marginal cost. However, since it is not competitive and, therefore, has relatively low elasticities, the price at which marginal revenue equals marginal cost will be much higher than the marginal costs.

The notion of charging all the market will bear also goes under the names of market-level or sustainable pricing of carrier access services. Basically, however, it means raising prices to the highest level that could exist in a competitive market without decreasing total revenue. But access service is not a competitive market in the normal sense of the word. Although uneconomic bypass is clearly a threat to a fraction of the access market, it is not so because the market is inherently competitive. It is so because the price of access is widely divergent from its cost, and this forces some end users to take actions that they would not otherwise take.

This is hardly the definition of a competitive market. Therefore, it is a foregone conclusion that market pricing of carrier access service will continue to produce unnecessarily high markups and cause economic efficiency losses and restrictions in the flow of information. It is the uneconomic subsidy markups themselves and the resulting uneconomic bypass that primarily cause the access market to appear to have competitive or elastic characteristics. Using this aberration to justify the use of market-based pricing is inappropriate. It will merely result in the substitution of uneconomic markups called market-based contributions for those called non–traffic-sensitive support.

If all uneconomic markups were eliminated, the fundamental inelastic nature of the access market would be restored. The use of market-based pricing under these conditions is equivalent to monopoly pricing and will lead to the same type of artificially inflated network usage rates that are causing today's problems.

It is interesting to note that uneconomic bypass causes some portions of the access market to appear to be elastic, and the mathematics of market pricing will often lead to short-term reductions in carrier access prices. If the access market were truly competitive, these same methods would produce access prices nearly identical to marginal costs. For these reasons, it may appear to be an attractive and reasonable procedure. But this is a dangerous illusion, for these methods can lead to the perpetuation of an economically inefficient telecommunications pricing structure in a market that is not fundamentally competitive.

Let me quickly add that market-based pricing is the lifeblood of true competition. Even though its implied markup can reduce economic efficiency, it is precisely this potentially large but transitory markup that fuels the engines of innovation. The great beauty of the competitive market is that it temporarily permits such markups as the incentive for excellence and the reward for success. However, its self-correcting mechanism ruthlessly seeks out and eliminates the markups through the pressures of competitive pricing. Any temporary losses in economic efficiency are more than offset by the constant kindling of innovative energy and its ally, technological progress.

Clearly, the access market of exchange carriers does not qualify for such a process. It is necessarily and extensively regulated, and most important, it possesses bottleneck facilities for the majority of information transactions. Therefore, the reasons for the existence of market-based prices in our society do not presently apply to that market and are not likely to apply to it in the near future. As a result, setting market-level prices for carrier access services will simply reduce economic efficiency. Although the losses may be somewhat lower than current levels, they will still be unjustifiable losses. Furthermore, they will continue to motivate uneconomic bypass, and they will perpetuate the ability to levy a tax on the flow of knowledge in our society.

Let me end this initial "tax protest" with a perfect example of a way to add insult to injury in the pricing of carrier access services. By using various types of capacity charges, some exchange carriers are now trying to reduce the market reaction to improper pricing, commonly known as uneconomic bypass. At the same time, they are advocating the use of market pricing for carrier access charges. Of course, the restriction of bypass activities will force the price elasticities of access service to appear lower than they otherwise would be, thus producing higher market-based markups.

With the capacity taxes in place, the exchange carriers can continue to impose high markups with relative impunity. At the same time, the capacity charges will cause the market-based markups to be larger than they otherwise would be. I call this interesting process the "gotcha" method of access pricing. Although I do not believe that the exchange carriers deliberately planned this trap, it is an interesting coincidence.

Unfortunately, because of the siren call of guaranteed subsidies from access services and the apparent lack of any strong convictions that uneconomic pricing is the real enemy, some exchange carriers are proceeding down this road. This is a sad turn of events for our industry and for our society.

I am hopeful that both the regulators and the exchange carriers will begin to focus on the long-term damage that improper access pricing will do to the flow of information in our society. They can do this by ensuring that economic pricing reforms are clearly underway before adopting any emergency and, one would hope, short-lived plans such as capacity charges. They can also remain true to the principles of economic pricing, which sanction market-based methods only when competitive market forces are present.

It is vital for our industry and its regulators to stand together to convince a skeptical public that repealing the knowledge tax is critically important to our society. Unfortunately, the general public and their advocates seem to be focusing only on potential increases in local telephone rates, and much of our industry is focusing only on preserving the revenue flow to the exchange carriers.

These priorities are rational and understandable, but I remain hopeful that the fear of losing the renaissance promised by the information age may persuade all sides to confront and eliminate the real problem, which is grossly distorted telecommunications prices. If this is done slowly, carefully, and completely, increasing numbers of people will be able to use the knowledge network fully. This is the real promise of the information age!

7
Cross-Subsidization in a Competitive Environment

Lawrence R. Weber
AT&T

This chapter will focus on a subject perhaps a little unusual for a representative of AT&T—local-exchange company (LEC) revenue streams. My interest is probably obvious. In the past year, my firm has been asked to make up many LEC revenue shortfalls.

Background

In the nine states across the Southeast, reaction to the January 1, 1984, restructuring of this industry was generally very similar, and regulatory proceedings followed a fairly consistent pattern:

1. Reserving the intra-LATA toll market to the local companies.
2. Establishing carrier access charges at levels generating LEC revenues equal to the support previously derived from inter-LATA tolls.
3. Extending discounts on access charges to OCCs ranging from 35 to 55 percent.

Regulators hoped that this approach would permit continued monopoly-style pricing policies based on subsidies between services. The mathematics, when worked out in the quiet of the hearing room, were impeccable. However, it is now apparent that these carefully assembled equations are jeopardized, if not doomed, by their failure to recognize the level of competition in both inter- and intra-LATA toll markets, as well as in the access market.

Regulators and local companies felt that the revenue flow to LECs from their intra-LATA toll services was secure. AT&T was uniformly restricted to inter-LATA service, and the perception was that this exclusion ensured the intra-LATA LEC revenue flow. However, it is becoming increasingly clear that despite the exit of AT&T from this market, such confidence is not justified. Inroads by resellers and OCCs are rapidly eroding local company intra-LATA revenues. Some of this erosion is consistent with regulatory directives, but some of it is simply a blatant disregard for commission authority.

Regulatory Loopholes In the Intra-LATA Market

Resellers—carriers that have elected to lease rather than purchase facilities—have almost uniformly been exempt from regulation. More important, they have been given intra-LATA operating authority.

In the nine southeastern states, there are eighty-three lease-based carriers competing with local companies for the intra-LATA business. Although these carriers have generally been perceived to be supplemental carriers, requiring little or no attention from regulators, their success in the market makes it clear that they play more than a supplemental role. Their entry into the equal access sweepstakes makes it even more clear that they are full-fledged competitors. Therefore, it is important that the type of regulation and operating authority resellers now have to be reevaluated in light of their new role and their effect on local company revenues.

To the extent that resellers capture business in the intra-LATA market—or, for that matter, the inter-LATA market—revenue flow to the local company is reduced. These carriers, because they lease network capacity, are now exempt from paying access charges on completed calls. Although the payments they make for facilities generally include some increment for access, the level of revenue flowing to the local company is significantly below the level received if access charges were paid on a per-message basis. This differential, inescapably, is recovered by additional charges either to local customers or to toll customers of other carriers.

The activities of resellers appear to be consistent with the decisions of regulators, but there is also widespread improper provision of interexchange service in violation of state commission orders. Carriers that are not certified to operate by state commissions are completing intrastate calls; carriers that are restricted to inter-LATA business are handling intra-LATA calls. Usage properly assignable to the intrastate jurisdiction, therefore, is instead credited to the intrastate jurisdiction. This creates an LEC intrastate revenue shortfall. Moreover, since interstate access charges are generally less than intrastate charges, this improper reporting results in reduced overall revenues for the local company.

The result is that some customers end up paying more so that others pay less. The shortfall probably means either increased access charges to the authorized intrastate carrier and its customers or some form of higher local rates.

Statewide operating authority and limited access payments by resellers, as well as unauthorized intra-LATA call handling by uncertified carriers, all help erode the intra-LATA toll and inter-LATA access charge revenue flow to LECs that was anticipated by regulators. Simply put, local company tariffs are being violated. Continued acceptance by local companies and commissions of acknowledged, improper reporting is contrary to the interests of those companies, the consumer, and other long-distance carriers.

In addition, some carriers have taken steps to participate in the equal access process despite the fact that many have not even applied for, let alone received, commission authority to operate. Responsible action on the part of local companies and state regulators is urgently needed to regain control of the situation. An analysis of OCC billing records would provide a basis for retroactive collection of lost LEC intrastate access revenues and a corresponding adjustment to collections from other carriers.

Regulatory Loopholes in the Inter-LATA Market

In the inter-LATA market, the current rules of competition also erode anticipated revenue flow to LECs. Discounted access charges to competitors other than AT&T are the reason.

Statistically valid studies indicate that where OCCs operate, they now serve as much as 34 percent of the intrastate long-distance business market and nearly 60 percent of that market's most lucrative customers. Clearly, this has serious consequences for AT&T. However, the implications for local companies are even more severe.

Because of discounted access given to all carriers except AT&T, there is a reduction in net income suffered by the local company when a call leaves AT&T and moves to a competitor. The local company's loss is greater than the income reduction suffered by AT&T. In fact, it appears likely that OCC access may be provided by local companies at a loss. This revenue shortfall is significant, and, ironically, it results from a discount structure the LECs themselves advocate. It is unfair to expect local-exchange customers or AT&T's customers—the majority of whom live in areas that have been ignored by OCCs—to make up the shortfall resulting from the discounted access charges advocated by local companies.

Added data regarding the extent of competition in the long-distance market underscores the impact of discounted access on local company revenues. More than 400 carriers now pay reduced rates for access. (AT&T, the one carrier that pays the full rate for access, has about 57 percent of the market.) Competing carriers' revenues approached $4 billion in 1983 and are growing rapidly. Clearly, the provision of discounted access on this volume of business reduces LEC revenue substantially. The public interest would be better served by pricing structures that not only treat customers equally but also maximize revenue support to local service rather than settling for a half-loaf.

Competition in the Access Market

AT&T is not a competitor in the access market; we prefer to be a customer as long as possible. However, numerous competitors are vying with the local company for this revenue, and there is an ever-growing body of evidence that local companies are, in fact, losing this battle. The growth of private transmission systems continues. In the nine southern states, more than 298 such systems are now in use. Surveys regularly confirm Touche Ross's findings that more than 50 percent of the largest business customers surveyed either now bypass or have plans to bypass. The reasons behind this trend are fairly clear. A study entered in a Florida access charge docket in August 1984 indicated that by using off-the-shelf hardware, large customers can reduce their access costs 80 percent.

The leaders in the movement to bypass are state governments, universities, banks, energy companies, and, yes, telephone companies. These groups all have strong trade associations or professional groups to manage or coordinate their efforts. Without significant changes in pricing and regulatory policy, local companies will continue to sustain significant reductions in access revenue streams.

The point of this discussion of competition and its effect on local company revenues is very simple. AT&T does not establish the regulatory framework of this industry. It does not price local company services, and it does not enforce certification and local company tariff provisions. AT&T cannot be held responsible, therefore, for the erosion of LEC revenues. Yet such thinking still prevails. A strong effort is now being made in one of our states by a group of local-exchange companies to get the state commission to access AT&T for those companies' revenue deficiencies in intra-LATA toll. This is being done despite the fact that AT&T has been excluded from that market and has steadfastly adhered to that exclusion.

Where Do We Go from Here?

Every time an AT&T long-distance competitor finds a new way to save its customers money, it dips into the local-exchange company revenue stream. That stream, however, must be replenished, and regulators are forced into the unenviable position of replenishing it by one of two means: higher access charges to AT&T and its customers or higher local rates. Regulators are thus forced into a no-win situation. Either rural and small town long-distance users must pay more to subsidize competition in large cities, or local rates must be raised for everyone.

Regulation is often no longer setting the standards for the industry but merely reacting to the competitive marketplace. It then becomes the industry waterboy, searching for new revenue sources to replenish the local revenue stream. Far-sighted regulation can adapt to this situation and, once again, take the lead in assuring equal treatment and maximum benefits for all customers.

In fairness to all customers—long-distance and local alike—the competitive marketplace must be allowed to operate more and more freely in today's telecommunications world. Evidence exists that it is already forcing its wishes on the still-regulated industry. However the transition evolves, prices of each service will be driven to cost and, more important, all customers will receive equal and, I believe, substantial long-term benefits.

8
Access Charges and the OCCs

James D. Ferguson
MCI Southeast

Under the best circumstances, it is difficult to get three people in the telecommunications industry to agree on anything these days. When the subject is access charges, it is impossible. But the access charge issue, as many other issues with which we are grappling in the regulatory arena, is the result of a major transition that is taking place in the industry—the transition to competition. In effect, what we are doing in our research is defining the systems necessary for our companies to operate in a competitive environment. It is no easy task.

The job is further complicated when there is a transition phase in which some players are beset by disadvantages that do not affect the others. The inability to deal with these problems in an equitable fashion threatens the development of the very competition toward which we are all striving. One of MCI's primary regulatory concerns is the provision of an adequate intrastate access charge differential during the transition phase. The Feature Group A interconnections provided to the OCCs by Bell are inferior to the Feature Group C connections given to AT&T. Consequently, we are allowed a 55 percent differential by the FCC on interstate access, and we had hoped for some consideration of this standard by the local public service commissions.

This differential is designed to compensate the OCCs for three competitive disadvantages inherent in the unequal access to which we are restricted throughout much of the country. First, the inferior access associated with Feature Group A connections forces the OCCs to incur costs that AT&T does not have to bear. Second, inferior access forces the OCCs to offer services perceived as less valuable to customers because of the inconvenience in making a Feature Group A call. And third, unequal access significantly limits the total pool of customers the OCCs can serve.

Because of inferior interconnections, OCC switches must spend more time actually switching a call than do AT&T switches. Unlike AT&T, the OCCs must also use their switches to collect and verify billing information at the time each call is placed. Both activities use switch capacity and increase switch costs. The OCCs also must install echo suppressors on their transmission links to improve the quality of signals degraded by interconnections. AT&T does not face the same financial drain.

Furthermore, the OCCs incur considerable expense compensating for such shortcomings as the absence of Automatic Number Identification (ANI) and answer supervision. The lack of ANI has made our network more susceptible to fraud than AT&T's network is. Since it is impossible for us to determine the number of the calling telephone, we must rely on authorization codes for billing. We cannot determine the validity of a subscriber's complaint that his authorization code has been misused, and we therefore experience a higher rate of uncollectibles—ranging as high as 10 percent for MCI versus less than 2 percent for AT&T. Also, since answer supervision is not passed to the line side of a local office, it is unavailable with Feature Group A interconnections. Consequently, we have been forced to invest in expensive computer systems to compensate for this loss.

Our Feature Group A customers have to dial twenty-two digits to make a long distance call. In addition, they experience quality loss that runs 12 to 13 decibels, versus only 6 decibels for a Feature Group C call. As you can imagine, these inadequacies have proved costly from both a technical and a marketing perspective.

The drawbacks to our marketing efforts do not stop here. Because line connections such as Feature Group A cannot accommodate rotary phones, the OCCs have been locked out of 50 percent of the long-distance market.

All of these disadvantages combine to create serious barriers to an OCC's ability to offer service. They also hamper the development of a truly competitive telecommunications market, without which the benefits of new services, lower rates, and innovative technologies will not exist. Realizing this threat, the FCC devised the access charge differential to compensate for our competitive disadvantage and to ensure service through the transition phase. Unfortunately, that is not the case with many of the state public service commissions. They do not yet understand the socio-economic movement that is taking place. They do not seem to realize that the transition to equal access will determine the development of the competitive environment necessary for society's progression into the information age.

Nationwide, the OCCs are entering intrastate markets for the first time. They are building new facilities and expanding their networks to offer new services and to meet increased demand generated by the expansion of information technologies. These carriers are creating the telecommunications infrastructure that will ensure the future, long-term growth of each state's economy.

The development of state-of-the-art telecommunications networks throughout the Southeast is becoming as important to the business community and the public in general as accessibility to airports, electrical power, and railways has been in the past. At this very moment, telecommunications is climbing the priority list of services necessary for a business to operate most efficiently—to have the competitive edge it needs to succeed.

Telecommunications is not only the highway *to* the future, it is the highway *of* the future in the information age. And without competitive telecommunications markets within the states, the infrastructure necessary for future economic development will not exist.

So far, however, only lip service has been paid in support of intrastate competition. The states have set various differentials for the OCCs; all, except Mississippi and Kentucky, are lower than the 55 percent implemented by the FCC, and in the case of Tennessee, there are none at all. In other words, the OCCs are being required to pay premium rates for inferior access. This failure to provide adequate differentials distorts the competitive process in favor of AT&T and will interfere with the development of competitive markets for years to come.

As equal access is beginning to roll out around the Southeast, MCI is expanding its network to serve those additional markets. We are spending over $1 billion this year alone to offer service to places like Houma, Louisiana, Boone, North Carolina, and Pascagoula, Mississippi, which we have been unable to reach before. At the same time, we are, in effect, having to keep two networks going to continue to serve our Feature Group A customers, who are connected to end offices that are not equipped for Feature Group D service. This amounts to a duplication of our network, which is yet another expense AT&T has not had to bear. In addition, we are virtually being penalized for serving our original customers by the absence of an access charge differential to redress the restrictions of lower quality interconnections.

The reasoning seems to be that the OCCs could purchase another type of interconnection to get them through the transition phase—or, if not, it really will not be too long before equal access is available everywhere anyway, so how much damage could be done? My answer is, A lot.

At this point, only AT&T is allowed to enjoy the premium interconnection of Feature Group C. It has never tried to obtain either Feature Group A or B on both originating and terminating ends. It has suggested, however, that the OCCs utilize Feature Group B-direct to realize some of the premium features from which AT&T benefits.

There are some good reasons that we do not. Feature Group B-direct is available only from Western Electric 1ESS and 1AESS switches. For other class 5 switches, the OCCs can now obtain only Feature Group A or Feature Group B-tandem, neither of which provides ANI.

Next, Feature Group B-direct would require the OCCs to construct direct trunks to every class 5 office on which such access was ordered. Because of our low traffic volumes by end office, each of those trunk groups would have only a few circuits. The OCCs now typically serve a metropolitan area through a single trunk group and therefore gain efficiency in combining traffic onto large trunks. If we were required to splinter traffic by using Feature Group B-direct, we would lose that efficiency. Moreover, every additional circuit would require an additional switching port. In fact, we would have to approximately double our switching and trunking facilities to employ Feature Group B-direct. Finally, ANI is transmitted in a different manner under Feature Group B-direct than under Feature Group D. If an OCC attempted to convert to Feature Group B-direct and then to Feature Group D, it would have to reprogram its switches twice.

So we are stuck with Feature Group A connections and yet we are being required to pay premium rates for access. This acts as a barrier to growth for the OCCs and impedes the progress of competition. Of course, AT&T does not mind. By controlling premium access but not paying rates that reflect its current scarcity, AT&T is able to protect and even expand its dominant position in the market at the expense of its competitors.

Why? Because the cost of entering and serving a market is heightened for the OCCs. Therefore, alternative carriers not only incur capital expense but are unable to offset the lower quality of their service by offering greater savings to their customers. Conversely, a properly calculated differential would permit competition during this transition to equal access by allowing the OCCs to charge more competitive intrastate rates despite the disadvantage of unequal access.

The bottom line is that I believe we are going to have to look carefully at the way access charges are set, both now and in the future. They must reflect the value of the service, technological efficiency, and cost-based pricing. Otherwise, the ability of the interexchange carrier to operate in many areas of the country will be jeopardized.

The OCCs today are restricted in the areas they serve by the time and costs involved in building their networks. As did the local telephone companies before us, we have initiated service in metropolitan areas first. At MCI, we have spent over $4 billion to be able to offer originating service in 450 cities. Our goal is to provide service to as many cities and towns as our network expansion permits—and that includes rural areas.

As you know, the antitrust division has postulated that approximately 100,000 access lines are necessary to support competitive entry into a market. Therefore, new networking arrangements may be required for the OCCs to serve rural areas after the local operating companies provide equal access. The establishment of methods by which traffic is aggregated sufficiently to allow service to a wide area from selected points within that area would help ensure that rural customers have access to competitive carriers—unless, of course, the OCCs are burdened by access charges that have no cost basis.

Dedication to similar technological efficiencies and access charges that reflect them will also reduce the possibility of bypass by large commercial users in metropolitan areas. If access rates are not cost-justified, neither the local-exchange carrier nor the interexchange carrier can control the loss of these customers. This technology is already available for large users to avoid artificially inflated communication costs. Bypass itself is not new. It has been utilized by large financial organizations, Fortune 500 distributors and manufacturers, aerospace firms, large educational institutions, and government agencies.

It is only the threat of bypass to the local-exchange company that is new. As we move into the information age, the ability to communicate and to move data is an increasingly crucial aspect of a business's ability to compete. It will become even more so. Consequently, it would only be practical for these companies to bring in-house the systems necessary to manage their communications needs if the price of service from the local carrier is driven upward by unrealistic access charges.

As the regional operating companies venture into new enterprises, such as providing information services, the potential for the old practice of cross-subsidization will again exist. And if access charges are unduly used to this end, the employment of bypass will flourish.

As I said earlier, this period of transition from a monopoly to a competitive world must be handled very carefully. We all have something at stake, and we all have a lot to lose. For the OCCs, intrastate access charges that fail to account for the differences in the value of and rights to premium local-exchange access will interfere with the development of competitive markets for years to come. For the customers, the benefits of competition will not be realized if the operating companies neglect the implementation of technological efficiencies and cost-justified access charges that will prevent higher rates. And for both the interexchange and local-exchange carriers, we will risk the perpetuation and growth of our customer bases if access charges do not reflect the value and cost of the service itself. Furthermore, we will stifle the advent of competition and lose step with society's progression toward the information age—a movement that will go on without us.

9
Access Charges: A Local-Exchange Carrier's Perspective

Alan K. Price
Bell South Services

Although my main task is to indicate how a local-exchange carrier views access charges, I must admit at the outset that some of what you will read will be my personal views. This book also gives the reader a chance to read what two of my biggest customers have to say about how *they* view access charges.

The telecommunications industry had rolled along for years and years in a monopolistic environment in which the prices of services were only very loosely tied to the costs of providing those services. Even when technology changes started to make some portions of our business much more profitable than others, we had little incentive to realign our prices, because realignment would have resulted in increases in local-exchange rates at a time when the industry and the regulators were pushing hard for universal service.

The initial tariff development effort (I was a member of the task force) was most difficult, because a way had to be found to design a tariff that replaced the internal accounting process used by AT&T to cover the costs and share the revenues that resulted from the interstate toll venture, while also replacing the interim tariffs used to provide services to the other common carriers. If that were not a large enough challenge, the stringent requirements of the Modified Final Judgment (MFJ) had to be satisfied simultaneously.

As you can imagine, there was a lot of give and take during this process, and because of all the stakeholders involved, a great many compromises were reached. All in all, in my opinion, the initial tariff produced by this task force and presented to regulators represented a very reasonable and equitable balancing of those varied interests.

The rest is history, so to speak. The tariffs underwent some of the most extensive investigation ever made by the FCC and state regulatory authorities. In addition, Congress and consumer organizations of all sorts became involved in varying degrees.

The tariffs that resulted from this process did reflect the interests and concerns of the various parties involved; however, they unfortunately did not adequately reflect the new competitive arena that is before us. The FCC, which initially laid out a tariff and transition plan that, in my opinion, would have fostered the growth of competition, was forced by pressures from other stakeholders to back down to such an extent that problems are now starting to surface. These problems may be just the tip of the iceberg unless regulators take measures that will position the provision of access services more in line with the original FCC plan.

The initial FCC access charge plan was designed to promote competition, prevent bypass, and ensure the continued availability of affordable local-exchange service. And even though I can not say that everyone agreed with all aspects of that plan, I do believe that had it been placed in effect, many of the problems coming up today would not have occured.

Consider bypass, for example. The FCC's plan would have provided a clear signal to interexchange carriers that the billions of dollars of subsidy built into the initial tariff rates would be phased out over a relatively short period of time, making it disadvantageous to them to try to avoid such charges in the short run. As it is, the new interexchange carriers are forced to consider bypass to maintain price differentials between themselves and the dominant carrier, and the dominant carrier is forced to consider bypass to compete with such differentials. It is a vicious circle, in which the more the competing carriers are successful in keeping their customers through such tactics, the worse off the local-exchange carriers are. Although it is hard to blame the carriers for doing whatever they deem necessary to keep their costs down and to enhance their competitive posture, the LECs certainly do not want to be in a position of having to pay the bill for such activities through either local-exchange customers or their stockholders.

It is my belief that a controlled transition of the subsidy away from the interexchange carriers and onto local-exchange customers would best accomplish this. I do not mean that we should just blindly decrease carrier charges and increase local rates. But I do believe that we should establish a transition plan that would accomplish just that. In the meantime, I believe we should begin looking at the various markets in which access services are utilized and determine the market value that should properly be associated with those individual markets. In other words, certain markets are more likely to support a higher charge for access than others.

Today, the existing tariffs require that every minute of use be assessed at the same rate, regardless of the volume of calls of a particular customer, regardless of the time of day, regardless of the direction of the call, and so forth. Individual market characteristics associated with the use of our access services are somewhat ignored in the development of charges, and as a result, carriers are encouraged to make decisions they otherwise might not make.

To remedy this situation, I offer two suggestions. First, I believe that the FCC, in concert with the joint board, should change its rules so that the majority of costs—and I emphasize the word *costs*—associated with the provision of a subscriber's local loop are removed from the interstate arena and, as a result, are no longer required to be recovered through interstate charges.

Second, I believe that local-exchange service has been extremely undervalued in the past and, therefore, that the rates for local-exchange service not only can be but will have to be increased somewhat in a competitive environment. I also believe that such increases, if placed in the hands of state regulators, can be accomplished without adversely affecting universal service.

I do not believe that just because such costs are assigned to the intrastate arena, they should be recovered totally from local-exchange rates, nor do I believe that they should be recovered totally from intrastate charges. Local-exchange carriers must be allowed to establish intrastate and interstate rates at market levels. In a competitive marketplace, different services are able to achieve different levels of profit. Although we, as a rate-base regulated company, are only allowed to earn a set rate of return overall, there is no reason that we should not be allowed to price services according to their value in the marketplace. If that results, as I believe it will, in revenues that can be used to support some segment of local-exchange services, then so be it. This concept is not new—in the past it was called residual pricing. Local-exchange carriers used this concept to keep basic local-exchange rates as low as possible. The difference between my views and this old concept is that local-exchange carriers should also be allowed to set the prices for some segments of the local-exchange market according to their value in the marketplace, thereby minimizing the subsidy that is needed.

But what does this all mean to interexchange carriers? It means that local-exchange carriers are going to have to get a lot smarter in the way they price services for their different market segments. It also means that LECs are going to have to find some way to convince state and federal regulators that they absolutely must have this type of flexibility in pricing if competition is going to grow and mature and if they are going to grow with it.

The predominance of LEC business is associated with the use of their local network. They have been building local network facilities for over 100 years, and they think they are pretty good at it. A huge amount of the LEC revenue stream is associated with the use of that network. As a result, they fully believe that they and our customers are better off if that network is utilized efficiently and profitably. To do this, LECs must simultaneously address the needs of all customers, regardless of their size and usage characteristics. They must offer services that compete with state-of-the-art bypass facilities for their very largest customers, and they must find ways for both them and the interexchange carriers to profit from the occasional use of very low-volume customers.

In short, these interexchange carriers are the LECs' biggest customers, and the LECs want their business. The LECs want them to feel confident that they can use their network to provide their services in a profitable manner. It is essential to their long-term viability.

10
Communications in the New Era

Frank Skinner
Southern Bell

Since divestiture, the question frequently asked of BOCs is, "Well, how are things going now that the Bell System has been broken up and you are separated from AT&T?" The fact that customers raise such questions perhaps suggests that, to this point, they have seen no real change aside from more pages in the phone bill. But the question also suggests a misplaced focus, because important as the breakup of the Bell System was, the change from a BOC's past as part of the Bell System to its future as part of one of the regional holding companies is far less significant than the change we are undergoing from a past of regulation to a future of competition. Right now, the BOCs are somewhere in between—moving rapidly into a new and vastly different era and heading into a future that will be more and more dominated by competition.

Behind this trend, of course, is a national policy consensus that favors competition as the way to meet public needs—a consensus that found formal expression in the FCC's Computer Inquiry II and in the Modified Final Judgment of 1982. Although the eyes of the world have been on the implications of the industry's restructuring for the long-distance market, it is quite clear that the move toward competition is on in the local-service arena as well—and in virtually every other service area. In fact, in looking over a BOC's monthly financial statements line by line, one would find it difficult to isolate any source of a BOC's revenues that is not facing stiff competitive pressure. This pressure is likely to increase, so even after divestiture, the BOCs are still a business in transition—and that is the way it is going to be for a long time to come.

The change in national policy and the continuing competitive evolution it has produced have major implications, not only for those who manage the telephone business but also for those who regulate it. The new era demands that the LECs find creative ways of ensuring excellent service to the public at the lowest reasonable cost, and that, of course, is not a new challenge. Just as the LECs were called upon previously to achieve universal service, so they are now called upon to maintain it in the new marketplace.

That is not all. We will also need to determine how that marketplace can best work, fairly and equitably, in the interests of companies and consumers alike. It is this challenge I would like to address in this chapter.

Year One

Considering the scope of the task the BOCs had to tackle, I think it is fair to say that they have already accomplished a great deal. Together, regulators and the industry have carried out a massive division of responsibility for the sale of equipment and enhanced services. The BOCs implemented the largest corporate reorganization on record, including a dramatic change in the way toll service is provided. And they have begun the tremendously complex job of repricing the industry's services to accommodate the changes taking place within the industry itself.

I wish I could say the BOCs have made this gigantic transformation without blemish—but of course that is not so. Perhaps the most notable—and widely publicized—concern we have had has been the problem of delays in filling special orders for business, orders that involve both BOCs and AT&T. Fortunately, this problem appears to have been substantially eliminated. Working together with AT&T, the BOCs have virtually eliminated the backlog of orders for special services that arose in the wake of divesiture. With those provisioning difficulties out of the way, it appears that the BOCs can now get on with the rest of their business.

Computer Inquiry II, divestiture, and the monopoly world they buried are prologue now. The BOCs are approaching the end of year one of the new marketplace, and I think it is important to ask what they want to say they have accomplished at the end of year two.

Year Two and Beyond

I would suggest that the BOCs still have miles to go. As a matter of fact, a highly definable agenda of unfinished business faces the telecommunications industry and those who regulate it—an agenda that will not be easy to carry out and that will require a change in how we think about the provision of service, about pricing, and about regulation. It is an agenda, however, that in the mutual interests of both the industry and its customers, must be addressed.

The first item is equal access. The LECs have to fulfill their responsibility to ensure that the bulk of their customers have the opportunity to use any long-distance company on an equal basis with any other. Southern Bell carried out the first of these so-called equal access conversions in August 1984, in Atlanta. It had 10 percent of its access lines converted at the end of 1984, and by the end of 1985, it had converted 70 percent. Meeting commitments in this area has been a top priority.

The second issue is the need to move steadily forward in attempting to devise pricing options that will meet the demands of the marketplace. An LEC cannot do that in isolation but, rather, must rely heavily on the industry-regulator partnership that has already accomplished so much for the consumer.

The competitive era of telecommunications is one in which the consumer increasingly will seek choices—choices in equipment and services and choices in how and how much to pay for those products and services. There are still strong pockets of resistance in some places to measured service options for customers who would like to take advantage of them. Particularly in an era when the flat-rate price will have to move ever closer to cost, it seems to some a disservice to tell consumers that they cannot have options that will keep their rates lower, even if they would like them. Conversely, to be instrumental in placing on some customer's table a wide array of information-age choices seems to be an accomplishment an LEC could point to with pride.

The third agenda item also concerns pricing in the changed marketplace to recover the costs of network access. Some BOCs, such as Southern Bell, have taken the position that the correct price for network access should cover its costs. They believe that long-distance and equipment sales, rather than producing a subsidy for that access, should remain what they are—products and services that add value to telecommunications usage.

In the competitive marketplace, there are scores of examples of that principle. For instance:

1. Exxon does not help you pay for your car, but it provides products that add value to your automotive purchase.
2. Kodak does not help you buy a camera, but it provides you with choices of products you can use with the camera you bought.

So is it not reasonable to believe that, given options that keep access widely available, local service should be a value-added service that presents the consumer a marketbasket full of choices?

This matter of recovering the cost of network access has a real impact on LECs, because only by being allowed to price competitively in the competitive marketplace can LECs avoid the most serious immediate threat to the viability of the public telephone network bypass. The simple fact is that new technology and outdated social pricing make a deadly combination. And competitors, many of whom are also among an LEC's most valued customers, are finding it increasingly in their own economic interests to go around the public network. Bypass today is no longer a trend that can be denied. The evidence is massive and conclusive, although there are apparently some who still have trouble recognizing that it has become a fact of life in our industry.

The extent of the problem was recently demonstrated by a study conducted by Touche Ross among Southern Bell's largest business customers. It found that of the 333 customers responding, 23 percent were already using bypass technologies. Another 16 percent of these customers had plans to institute bypass systems in the immediate future. Cost was the major factor in a large majority of cases.

A close, unbiased look at such findings suggests strongly not only that bypass is no laughing matter but also that it may put universal service itself in danger. If the larger business customers who pay much of the freight for the network, are no longer on-line, low-volume customers will inevitably be left to support it. It may already be later than we think. Anyone who has ever seen an exponential curve knows that its greatest growth occurs in a very short time. By its very nature, bypass implies the possibility of sharp increases in the cost of telephone service. Once we reach a certain point on such a continuum, the process becomes virtually irreversible.

If regulators and operating telephone companies hold one value most deeply in common, it is a commitment to universal service. The trust the public places in those who regulate the telecommunications business is, to a large extent, based on their commitment to the maintenance of widely available service at affordable prices. At the same time, many BOCs have built their success, from their earliest days forward, squarely on the broad foundation stone of universal service. So it seems that there is considerable irony in the current question about the bypass issue. By responding to an understandable desire to see that rates remain low in the short run, LECs run a serious risk of endangering the very universal service concept they seek so devoutly to protect.

The key to universal service protection is protection against bypass. And the key to bypass protection is pricing flexibility—the ability to price services in line with costs and with current market conditions. Is universal service jeopardized by price increases of the sort currently being envisioned? If it is, we have certainly seen no evidence to that effect—and a multitude of studies have been devoted to this very question.

More realistically, is it not about time LECs turned their attention to an appropriate means of maintaining network access for that 1 or 2 percent of their customers who may find themselves unable to afford to pay the normal price for telephone service, even with price options? A wide variety of proposals have been suggested, and the LECs are apparently ready to work with regulators and legislators in devising an appropriate subsidy mechanism, if one is deemed needed, for that 1 or 2 percent. But given the likelihood that an equitable solution to that question can be found, it seems that the time has come to allow the great majority of the LECs' customers to pay their own way.

Finally, the fourth and perhaps most perplexing item on the unfinished agenda is the recovery of our capital investment at depreciation rates that are appropriate to an increasingly competitive industry.

The driving force behind the dramatic shifts in the telecommunications business can be identified in a single word: technology. A plethora of innovations since the 1940s and 1950s have spurred the telecommunications industry, from a trickle of would-be competitors in the late 1960s to the wide-open equipment and long-distance competition of today. And to the extent that bypassers are utilizing every kind of technology imaginable to avoid the local network, we can add local-exchange service to that same competitive category. The net result of these changes has been just what one would expect in a competitive market: reduced costs and greater choices for the customer. Technology has fed competition, which has, in turn, produced a fundamentally more efficient marketplace.

The changes already in place and still under way in the industry amount to a fundamental shift in the framework of human communication. It is a shift that is creating this whole new era many are calling the information age. The LECs are working hard to make these new opportunities available at the local level. From high-speed digital switching and fiber-optic cable transmission to local data transport and common channel signaling, the local network is being transformed. It is being made vastly more flexible and "intelligent" than ever before, in such a way that the cost of serving customers is being improved.

The meaning of all of this activity is that the LECs are increasingly well positioned to bring information-age services to their customers: services such as home shopping, home banking, and home energy management; services such as nuisance-call rejection and selective call forwarding; services such as enhanced 911 emergency dialing, which allows police and fire departments to determine automatically the location of the calling party. Technology is providing these opportunities. And the LECs are getting on with the job itself. They are mounting an aggressive modernization program aimed at making these and other new services possible—and at cutting our operating costs at the same time.

Most of an LEC's capital cost is met with funds internally generated through the capital recovery process. Yet the currently prescribed depreciation rates are simply inadequate to reflect the capital replacement realities of a competitive marketplace. For instance, fully 37 percent of the useful life of Southern Bell's plant and equipment has been consumed, but only about 18 percent of the asset value of that plant and equipment is accounted for in its depreciation reserves.

A statistical comparison of Southern Bell and eight of its major competitors was made recently. It showed that although Southern Bell compared favorably in many areas, it had the lowest asset turnover of the group, principally because its depreciation rates were so outdated. Depreciation, the study showed, was a major stumbling block in its way to being fully cost-competitive.

If depreciation rates were current today, Southern Bell's revenue requirements would be $500 million less than they currently are. That would obviously benefit everyone and allow Southern Bell to meet competition more effectively. Of course, bringing rates to a current level would involve somewhat higher access line charges in the short term, but because of the nature of the depreciation process, a crossover point would be reached in five years or less. From then on, rates would be lower than they would be under current depreciation schedules.

Technological change is moving ever more swiftly, and the opportunities for new services are ever more attractive. The LECs owe it to the ultimate beneficiary—the average consumer—to position themselves *now* to bring him or her those opportunities—to bring to the consumer the benefits the industry restructuring was designed to produce.

This, then, is the agenda the telecommunications industry would do well to face squarely today, so that it can say at the end of year two that it made great progress toward bringing the public into the information age. Effective provision of equal access, development of meaningful pricing options, recovery of the fixed costs of network access, and depreciation rates that reflect the competitive marketplace—that is a heavy program to say the least. But it is a fair program. It meets the responsibility of telecommunications as a business—to be competitive—and it would enable the regulators to protect the public interest.

Facing some of these challenges will take courage and determination, and I am not unmindful of the political process and the hazards of actions perceived by the electorate as costly. That is why it is so important to make it clear that telephone companies are handling each and every user of telecommunications service. They are bringing the users a world of video, voice, and data the likes of which they have not seen before. They are opening doors to choices of services that are giving users' communications ever-increasing value, making their hours more productive, increasing their feeling of security, and lowering their costs for long-distance calls and equipment. If the telecommunications industry meets these unfinished challenges, and as it improves and broadens the technology, it can say confidently to the public it serves that it is bringing them a better communications bargain than ever before.

11
Divestiture and the Local-Exchange Carrier

Ward H. White
United States Telephone Association

One of the primary goals of the United States Telephone Association (USTA) is to heighten awareness of the local-exhange carrier's role. That role has certainly become more complicated with divestiture. There is understandable confusion among the public about what has happened to the local telephone company. Even those in the industry, as well as the policymakers in Washington and the state capitals, have difficulty sorting out cause and effect and predicting the future with any degree of certainty.

The particular problem, of course, is that we are not dealing with a completed act. Divestiture is an ongoing process, and the rules that were established in the immediate aftermath are now being revisited. Also, the trend toward deregulation that preceded divestiture continues. Some parts of the industry have been deregulated, and there is movement in other areas toward that end, but we still must do business in a mixed competitive and regulated environment.

The local-exchange carrier industry is very much in a state of transition. However, we have seen a good deal of progress, so we are optimistic that a more final report will show local companies surviving, prospering, and continuing to excel in providing service to all customers.

One of the sources for optimism is knowing that, in a way, we have been here before. Today's $200 billion in plant and more than 114 million access lines were not put in place for us. They are, instead, the result of hard work and determination. And just as spirit was the driving force behind construction of the world's finest telephone system, it will guide whatever must be done to make current changes work for that system.

Any complete discussion of divestiture must begin with events long before January 1984. I am not going to attempt a thorough review here, but I do want to mention several occurrences that stand out in the chronology.

As early as 1930, in *Smith v. Illinois,* the Supreme Court decided that there should be a means of separating telephone property, revenues, and expenses between intrastate and interstate operations. The Court acknowledged the difficulty of allocating costs but accepted the general approximation of relative use, observing that "extreme nicety is not required." In the years since, the concept of interstate and intrastate, including local jurisdiction, has prevailed. But it now seems appropriate to be considerably "nicer" in applying it!

As technological innovations reduced the cost of providing long-distance service, accompanying rate reductions stimulated increased usage. Under existing separations procedures, this resulted in an increase in the assignment to toll or non–traffic-sensitive costs—or those associated with connecting a customer's phone to the central office—because of the use of a weighted formula known as the subscriber plant factor (SPF).

By the early 1980s, the SPF allocation to interstate of non–traffic-sensitive costs for Bell companies was three times greater than the measure of simple relative use. Such a subsidy-by-toll of local costs was acceptable in a monopoly environment, but in a competitive environment, such a subsidy will give large toll users the incentive to find or build alternative systems—and today's technology will make doing so possible.

Therefore, to halt the growth of that subsidy, the FCC froze the interstate factor at 1981 levels and, since January 1, 1986, requires an eight-year transition to a gross allocator of 25 percent for interstate non–traffic-sensitive costs. This gross allocator transition applies to all companies. This is one of the ways in which it is evident that what happens to Bell companies also happens to independents—and what this means to Bell customers it means all the more to the independents' customers. Independents have a smaller customer base and frequently serve rural or remote areas with higher costs per loop. To cover those costs, customers have to pay higher local rates.

A related focus of change is a pricing methodology for local and long-distance service. As long-distance subsidies are reduced, each part of telephone service must begin paying its own way. Part of the costs of the local loop that were previously recovered through long-distance rates are now recovered through a monthly charge on both residential and business customers. The FCC now requires telephone companies to collect a $6 per month charge from business customers with more than one line and a $1 per month charge from single-line business and residential customers.

All telephone companies providing local service must collect these charges, including, of course, some that distinguish these costs for interstate long-distance use of the subscriber's line from any increase in local rates.

For their part, long-distance companies have begun paying a per-minute access charge for their use of the local-exchange switching and trunking facilities that originate and terminate long-distance calls. Their payment for fixed or non–traffic-sensitive costs will be largely phased out (except for contributions to the Universal Service Fund) as the subscriber line charges begin to sustain the local network.

Subscriber line charges and interexchange carrier common-line charges are now funneled, in an accounting sense, to the National Exchange Carrier Association (NECA) for distribution to local carriers. The NECA is another by-product of divestiture, established by the FCC to prepare and file access charge tariffs and to administer the resulting revenue pools.

Naturally, another layer of management means more paperwork. In fact, one small rural company reports that whereas there were once six annual reports to prepare, there is now a minimum of ninety-five—to be done with the same eight employees who existed pre-divestiture.

Other changes involve customer premises equipment (CPE) and inside wiring. Because the FCC wanted to foster a competitive environment for CPE, it deregulated new CPE on January 1, 1983. There is no longer a recovery from the settlements process for new telephones, PBXs, and other CPE purchased or leased. At the same time, the FCC said that embedded CPE (those on the books as of December 31, 1982) would be subject to an allocation to toll that would be phased out over sixty months.

As in other areas, the FCC intent with inside wiring is to make the cost-causer the cost-payer. Therefore, new installations are now expensed, rather than capitalized, so that costs are borne by current ratepayers rather than by future ratepayers.

It can be noted here that bringing costs in line is only one of the issues that must be dealt with in these changing times. For example, although the FCC has now deregulated new CPE and is in the process of deregulating embedded CPE and inside wiring, two elements of shared-tenant systems, this policy has not been implemented uniformly. The Modified Final Judgment precludes Bell companies from competing for shared-tenant customers because they cannot offer least-cost routing for interexchange traffic. If shared-tenant and other services are to be offered by competitors of local-exchange carriers, it is inherently unfair to prevent Bell from doing the same.

Indeed, one of the most sweeping effects of divestiture and deregulation has been to inhibit competition. Barriers to entry are being eliminated, but local-exchange carriers continue to be bound by franchise obligation and rate-of-return regulation. Recently, the FCC issued an order partially granting Cox Cable's request to provide intrastate data services free of state certification requirements. This is another in a series of decisions by the FCC that will subject local-exchange companies to competition while they remain fully regulated.

Inequities, of course, were the core of one problem now being resolved through a painstaking process called equal access. Since the early 1970s, when MCI was, in theory, authorized to provide long-distance competition, there have been practical obstacles. AT&T, as owner rather than customer of much of the network, has had some obvious advantages. Concern over these advantages has grown as other competitors have entered the market. With divestiture, the Modified Final Judgment required that the seven new regional holding companies begin conversion of central office equipment to correct the inferior connections of AT&T's competitors. It is an enormous undertaking and thus was scheduled over a three-year period.

The timetable calls for all regionals to have completed conversion of exchanges with more than 10,000 access lines by September 1986. GTE is to convert two-thirds of its offices, which serve more than 10,000 access lines, by September 1987 and the remaining one-third by 1990.

The FCC ruled that independent phone companies are required to convert stored-program central offices to equal access within three years after receiving a "reasonable" request. The commission will establish a waiver process to handle those cases in which conversion would impose serious hardship on an independent company.

Some independents have voluntarily embarked on conversion programs. Others will probably follow suit to provide their customers with like connections to whatever long-distance carrier they choose. This is another example that shows that what happens to Bell companies happens to independents.

All local companies, Bell and non-Bell, must follow a ballot and allocation process as equal access reaches their exchanges. Customers who do not return ballots with a choice of long-distance carrier will be allocated to one according to the percentages of customers who do express a choice.

Neither equal access nor the ballot and allocation process will be implemented without a certain number of problems and, unfortunately, some more customer confusion. However, local-exchange carriers are cognizant of this and are doing as much as possible to ease the effects for the industry and for customers.

The USTA helps in these efforts. Besides representing the local-exchange carrier industry before policymaking bodies, the USTA also contributes to public relations on behalf of the industry. Currently, the association is engaged in a national information program—a concerted effort to explain to the public what is happening to its telephone service.

Divestiture has reached out to touch not only the ways we do business, but the ways we account for how we do it. The revision of the Uniform System of Accounts (USOA) provides the classic example. The FCC has promulgated an order which we hope will maximize continuity of data, minimize implementation costs, and retain the USOA as a financial reporting system, not a service costing system.

Specifically, the USTA has maintained that the proposal for an extensive new numbering system be set aside as ill-timed and overreaching. The USTA does acknowledge the need for some other changes—those that incorporate generally accepted accounting principles and that alter current capitalization practices to expense more and capitalize less. These positions, by the way, are a distillation of the virtual deluge of comments the FCC received regarding its proposed rulemaking. More than 160 parties responded with comments, many of them small companies that had never before participated in a commission proceeding.

This, too, must be pointed out as one of the happier circumstances accompanying divestiture. There is a depth of industry concern, a willingness to express it, and a meaningful way of doing so through the collective voice the local-exchange carriers have found.

The implementation of subscriber line charges I mentioned earlier is, to a large extent, the result of an industry unity agreement that took congressional pressure off the FCC and allowed it to go forward. That unified industry position also called for a Universal Service Fund, which will help high-cost companies keep customers on the public network.

Beginnings—there seems to be an abundance of them; certainly, they are easier to identify than ends are to predict. I have touched on only a few of the aspects of divestiture, and I hope you have noted that the dates associated with many precede by a number of years the historic headlines announcing the breakup of Ma Bell. I hope you have noted as well that the continuing tidal waves attendant to that event affect not just the mother ship and her spun-off parts but many other voyagers.

The future—that is a matter of perspective. Scientists and engineers talk of needs-based delivery of telecommunications systems to remote parts of the world. One can not help being excited about the possibilities for a global community. At the same time, there is the need to recognize that our own nation's network must be preserved to be part of it.

Toward that end, the USTA is trying to address the daily disruptions while at the same time taking the longer view. We have contributed to a study of the future undertaken by the National Telecommunications and Information Administration, an arm of the Commerce Department that advises the president. A first report from that study has now been released. Among other things, it notes that "foreign telecommunications policymakers will be reviewing the American experience, looking for documentation" and that "the focus on transitional problems has taken emphasis away from substantial gains achieved." The growth in the numnber of service providers, revenues, and user demand has been documented, according to the report.

The benefits, real and anticipated, are, after all, what the colossal changes are all about. But let there be no mistake—realization of the broad goals depends on solving some very serious problems. We believe there is a need for the following:

Reduced regulations and reporting requirements and increased pricing flexibility.

Fewer structural and jurisdictional separations that inhibit productivity.

A capital recovery policy that encourages technological innovation.

A merger of federal and state requirements into a coherent national policy.

With respect to some of these needs, I can report some encouraging signals. The FCC undertook what it called Computer Inquiry III. Computer Inquiry I, in the early 1970s, set distinctions between communications and computer industries, leaving the latter unregulated. Computer Inquiry II, in 1980, recognized that distinctions were no longer so clear. Policies were developed to separate carriers' basic services from enhanced services, and enhanced services were deregulated. But AT&T and the Bell companies, as dominant carriers, were required to make enhanced service offerings through separate subsidiaries.

The combined effect of the Computer Inquiry II definitions and the structural separations requirements has meant that the public is denied the ability to realize efficiencies that can result if enhanced services are integrated with carriers' basic services.

In Computer Inquiry III, the FCC will comprehensively reexamine these policies. Proposals include a redefinition of the types of activities a carrier undertakes, the elimination of strict structural separation in favor of lesser separation techniques, and the inclusion of a carrier's dominant or nondominant status in decision making.

In another key area, U.S. District Judge Harold Greene has opened arguments on the consent decree for the first time in an extended period. All motions relate in some way to a clarification of the exhange services permitted at the time of divestiture, including cellular radio, voice storage and retrieval of cellular messages, and shared telecommunications services.

In the area of taxes, this industry has made what we believe to be a strong case for being treated fairly in the tax reform process. Specifically, a panel of industry leaders has testified before the House Ways and Means Committee about the inequities of the proposed capital recovery system. As currently drafted, the proposal would require that certain electronic central office switching equipment be written off over a period that is 40 percent longer than the write-off period for computers. The new switches that phone companies are installing are computers and should be treated that way. They perform the same function as computers in any other industry and therefore should be subject to the same depreciation classification.

Other concerns for this capital-intensive industry include the proposed repeal of the investment tax credit and a proposed "windfall recapture" tax. The proposal addresses the normalization issue—the tax accounting principle that spreads out tax benefits for capital formation incentives over time to benefit current and future ratepayers—but it does not address the issue comprehensively enough to assure fair tax treatment for regulated utilities.

In a related area, the USTA will be filing an amicus brief with the Supreme Court to support federal preemption of depreciation rules and to introduce economic analysis that demonstrates the need for national policy on capital recovery.

Part III
Pricing Telecommunications in an Asymmetrically Regulated Market

Part II consists of six chapters with a common theme—that extensive subsidization among markets is untenable in the post-divestiture era. The title of the chapter by Robert T. Burns, "Separations of Costs and the Need for Reform," reflects its central theme and focus. Mr. Burns emphasizes that the purpose of separations is to assign costs, revenues, and taxes to the intrastate and interstate markets. It is a massive task (property values nationwide amount to $164 billion), but the joint-cost problem makes these allocations highly arbitrary. After reviewing the history of separations, Mr. Burns explains that the traditional allocations of fixed costs to the interstate carriers, combined with the practice of recovering fixed costs from usage-based tariffs, is untenable in the post-divestiture environment. Technology and newly allowed competition have combined to promote bypass whenever and wherever prices deviate from costs. Therefore, there is a need to reform the separations process. He also outlines the United States Telephone Association (USTA) objectives for separations reform and further argues that all of the stakeholders (not just AT&T, the FCC, and NARUC) need to be involved in designing the new rules as they evolve through negotiation.

Dr. James A. Leggette also deals with the issue of the historical evolution of separations in his chapter, and he traces how revenue requirements were shifted to interstate toll traffic. Dr. Leggette then discusses the economic policy issues that have resulted from loading non–traffic-sensitive (NTS) costs onto the traffic-sensitive toll market. He believes that the recovery of NTS costs through usage-sensitive rates is unsustainable in a competitive environment. In the final section of his chapter, he offers some recommendations for policy. He covers much of the same ground as Mr. Burns, though generally in somewhat more detail. However, both Dr. Leggette and Mr. Burns also reach the conclusion that uneconomic bypass is inevitable when prices deviate from costs. Dr. Leggette goes on to argue that universal service would not be threatened by cost-based pricing, because the demand is relatively inelastic. However, he allows that targeted low-income subscribers may have to be subsidized.

Dr. Frank Alessio emphasizes the shift toward cost (or market-based) pricing, deaveraging, unbundling, and market segmentation and the need for flexible pricing arrangements, eventually followed by deregulation. He believes that five principles should be kept in mind in establishing prices: (1) prices and costs are not synonymous; (2) they are both value concepts; (3) opportunity costs are not book costs; (4) nor are they historical or fully distributed costs; (5) but they are, most frequently, incremental, marginal, or avoidable costs. Each of these principles leads to a regulatory challenge that Dr. Alessio states concisely.

Jack Huber addresses post-divestiture pricing from the viewpoint of a local-exchange carrier (LEC). The challenge for the LEC is to manage a traditional and mature set of business activities, using common technological and administrative resources and adapting them to the emerging business opportunities that are present in a deregulated environment. He first reviews the conditions that prevailed in the averaged or subsidized world of pre-divestiture and contrasts that with the more highly differentiated post-divestiture situation. He feels that the more pressing issues under regulated competition are bypass, new marketing opportunities, and more efficient utilization of resources. Management's challenges are in the areas of research, product development, the velocity of change, and customer relations. He emphasizes both the problems and the challenges of the post-divestiture world.

In his chapter, Clinton Perkins recognizes the conflicts as well as the harmonies of interest among buyers and sellers of telecommunications services. Buyers want low prices—sellers want high ones. However, prices must cover company costs with a margin for profit. Mr. Perkins supports Dr. Alessio's principles of pricing as well as the idea that prices should reflect value to the customer, constrained by conditions of competition and underlying conditions of cost.

12
Separations of Costs and the Need for Reform

Robert T. Burns
Southern Bell

The theme of this chapter is the separations of telephone company costs in the changing world of telecommunications and the urgent need to reform the process. Separations is the telephone industry's version of financial cost accounting. The purpose of separations is to apportion or assign property costs, revenues, expenses, taxes, and reserves between the two regulatory jurisdictions in which telephone companies operate—intrastate and interstate. For BellSouth alone, this comes to a monthly apportionment of over $25 billion in property costs. The nationwide telecommunications investment base subject to separations is approximately $164 billion.

Telephone companies perform jurisdictional separations to coincide with our nation's regulatory structure. The Federal Communications Commission has regulatory authority over all interstate communications, and state governments regulate intrastate services through public utility commissions. Each government jurisdiction regulates different services, and rates for these different services are influenced by the overall revenue requirements, thus eliminating any ambiguity regarding what costs must be recovered from what jurisdiction.

But these jurisdictionally distinct services are typically provided with the same plant and equipment. For example, a local access line can be used to make both intrastate calls and interstate calls. Therefore, the investment in that line must be "separated" between the two jurisdictions for regulatory and ratemaking purposes. To understand how telephone companies currently perform separations, let us take a brief look at the history of separations.

The History of Separations

The history of separating plant and expenses between interstate and intrastate services for the purpose of establishing rates began with the Minnesota rate cases, which were decided by the U.S. Supreme Court in 1913. Although these cases involved railroad carriers doing interstate and intrastate business, they established the basic principle that property costs should be allocated to each jurisdiction according to the relative use made of the property. This basic principle remains in effect today, although political and social trade-offs have caused results somewhat different from what pure relative-use principles should produce.

In 1930, the Supreme Court case *Smith v. Illinois Bell Telephone Company* considered telephone industry separations for the first time. In that case, the court decided that separation of the interstate and intrastate property, revenues, and expense is essential to the appropriate recognition of the competent government authority in each field of regulation.

Until that time, separations had been performed on a toll-board-to-toll-board basis. This took into account only those costs associated with the use of facilities from toll board to toll board. Since the majority of telephone costs are related to facilities necessary to connect subscribers to the telephone network, this process significantly limited the levels of costs assigned to the interstate jurisdiction. Under the board-to-board concept, these local subscriber connection costs were recovered solely through local-exchange rates. When the 1947 *Separations Manual* introduced the station-to-station concept, virtually all of a telephone company's costs became subject to jurisdictional separations. This, of course, increased the total costs assigned to interstate.

Between 1950 and 1971, revisions were made to the *Manual* about every four or five years in the ongoing effort to maintain parity between interstate and intrastate rates and earnings. Each time the *Manual* was revised, a larger percentage of total costs was assigned to the interstate jurisdiction, so that all telephone users (exchange and toll) would share in the profits being generated by interstate toll service. These profits were accountable to the great growth in the demand for long-distance service and to economies of scale inherent in the nationwide long-distance network.

In 1971, the Ozark Plan was introduced. By introducing the subscriber plant factor (SPF) for the separation of non–traffic-sensitive subscriber plant, the Ozark Plan substantially changed the *Manual*. This factor remains in use today and causes about three and one-third times as much subscriber plant to be assigned to the interstate jurisdiction than would be justified by usage.

Although the subscriber plant factor was frozen in 1981, we are just now beginning to make the transition to the recovery of the geographically local exchange costs from interexchange users to end users.

Current Separations

It is important to recognize the most of the separations procedures in use today were developed in a noncompetitive era. Moreover, the conventional definitions used today have been greatly affected by delicate political, social, and industrial compromises, so that although the results of many allocations may appear reasonable and exact, they are, in fact, highly arbitrary. To better understand this, let us examine the allocation procedures for subscriber plant investment.

Beginning with the 1971 Ozark Plan, separations classified plant as either traffic-sensitive (TS) or non–traffic-sensitive (NTS). Traffic-sensitive costs are those that fluctuate with the use made of the facilities. They are, therefore, variable costs. Non–traffic-sensitive costs do not vary with the level of usage; they are thus considered fixed costs. Use of the SPF, as described earlier, results in an interstate assignment of more than three times the result of an assignment based on relative usage. For example, if 10 percent of a telephone company's total traffic is interstate, approximately 33 percent of its NTS plant would be assigned to interstate under SPF. As a result, high-volume toll users currently are assessed many times the cost of access through toll rates and are forced to look for bypass alternatives. In today's competitive environment, this excessive allocation threatens the very viability of the local exchange telephone companies.

This threat comes not only from the overassignment of costs to interstate but also from the usage-based rate mechanism for recovery of these costs. Inflated fixed costs recovered on a usage-sensitive basis result in a situation in which high-volume users of access services, such as AT&T, are charged many times the LEC's actual cost of providing the access service. To avoid these high-costs, interexchange carriers are beginning to use alternatives to LEC-provided switched access. Since the fixed costs stay with or without high usage, the remaining small customers and, ultimately, the local subscribers will be forced to fund the large NTS costs.

To the extent that LECs are required to load arbitrary costs (that is, costs not incurred in providing access) into their access offerings while competitors can avoid these charges, LECs will continue to face a significant competitive disadvantage. AT&T's MEGACOM offering is a prime example of exploitation of such costing and pricing inefficiencies. MEGACOM and MEGACOM 800 service will provide strong economic incentives for large users of WATS and 800 service to forgo switched-access arrangements and avoid the NTS subsidy.

Many events over the past several years have caused or aggravated the situation we find ourselves in today. The Department of Justice, in FCC order CC 85-197, stated that the Modified Final Judgment, which ordered AT&T's divestiture of the Bell Operating Companies, was not intended to disturb the jurisdictional separations process established by the Communications Act of 1934. Although the separations rules were not specifically altered as a result of the MFJ, it is now clear that the allocation procedures, which were reasonable in a nationwide monopolistic environment, are no longer appropriate today. Therefore, the MFJ will continue to necessitate changes to the separations procedures.

Rapid advancement in technology is also exerting pressure on the separations process. Technologies such as microwave and satellite transmission facilities reduce the sensitivity of costs to distance. Modern chip circuitry used in digital switching equipment blurs the cost distinction between toll and local processing.

Significantly, by 1990, BellSouth will serve one-third of its offices with digital switching machines. A separations anomaly related to digital switching is that it receives a higher NTS assignment than other technologies; therefore, interstate costs increase every time an office is added or replaced with the newer technology. This fact alone will cause BellSouth to shift $50 million of annual revenue requirements to interstate by 1990. In addition, many new services were not even contemplated by the *Separations Manual*—for example, packet switching, shared private network, and Integrated Services Digital Network (ISDN) or digital transport.

The Need for Separations Reform

Reform of the separations process is urgently needed for the following reasons:

Existing procedures are complex and voluminous and result in large operational costs to the telephone companies.

New technologies and post-divestiture operations are not recognized by the existing *Separations Manual.*

Local-exchange companies are experiencing pressures on their ability to price competitively, since the present separations procedures drive the complex access-pricing rules.

Finally, rules that govern the telecommunications marketplace have changed. Therefore, public policy considerations of the precompetitive era must be thoroughly reexamined in light of today's competitive environment.

The USTA

The United States Telephone Association (USTA) has established an industry committee to develop separations reform objectives as follows:

Separations procedures should be simple, explainable, flexible, and cost-effective, and should produce reasonable results.

Separations should achieve rational allocations of regulated costs without unnecessary cost loadings (subsidies), such as SPF, weighted toll minutes-of-use, or toll-contact factors.

Separations should only define what is state and what is interstate and should not dictate method of cost recovery.

Separations should not define costs to be deregulated for local-exchange company competitive services.

Principles should be uniformly applied but should not inhibit regulatory innovation.

High-cost concerns may be incorporated in the separations process for a limited period. In the long term, however, operations cost characteristics should not influence separations.

Separations principles should attempt to balance the interests of small and large exchange carriers as well as those of other stakeholders.

BellSouth is actively working with the USTA and regulatory commissions to develop and implement separations reform objectives. In addition, BellSouth and most other Bell Operating Companies are actively working to streamline and simplify procedures that do not require regulatory rule changes.

Achieving Reform

Prior to 1984, AT&T, the FCC and NARUC alone negotiated changes to the separations rules. Even when limited to these three major players, rule changes were difficult to achieve and very slow to be implemented. Today's telecommunications environment contains numerous stakeholders with diverse and often conflicting interests. One need only review the wide array of comments filed with the Joint Board concerning the narrow issue of end-user service order processing to appreciate the number and diversity of viewpoints. The major players in the game today include regional Bell companies, AT&T, independent telephone companies, other common carriers, resellers, the FCC, state public utility commissions, telecommunications customers, and numerous trade associations and consumer groups.

The key to achieving reform is to develop procedures that will balance the interests of these diverse groups. Trade-offs are inevitable, but workable changes are possible only if the trade-offs are shared equitably among all the conflicting interests.

Conclusion

The transformation of the industry brought about by rapid advancements in technology, divestiture of the Bell System, and competition has resulted in an urgent need to modify the industry's costing and pricing structure, which was constructed in a monopolistic environment. Today's business environment dictates a comprehensive overhaul of subsidies and artificial pricing constraints of the past in favor of economic costing and pricing. However appropriate the concepts and procedures of the past may have been, they are now obsolete and even economically detrimental. Because these procedures threaten the viability of the providers of telephone service, they also threaten the availability of the high-quality, reasonably priced telephone service that American consumers have traditionally enjoyed. For these important reasons, separations must be subjected to overhaul, simplification, and reform.

13
Separations and Telecommunications Pricing Issues

James A. Leggette
ATT Communications

The process known as jurisdictional separations and its relationship to recovery of non–traffic-sensitive (NTS) costs[1]—those costs that do not vary per call—are the key to understanding the need for reform of telecommunications pricing. Simply put, separations is the process of allocating costs between regulatory jurisdictions. At first glance, it would appear that separations is an area of interest only to cost accountants and engineers, but since it is used by the state and federal regulatory commissions to determine revenue requirements and, ultimately, price, it is of interest to economists as well. The stakes are not small: in 1982, interstate toll service provided at least $9.8 billion[2] to the local telephone companies to cover NTS costs.

Because separations and rate making are so tightly entwined, separations is at the heart of a variety of issues confronting policymakers. Among these issues are access charges, cross-subsidization, universal service, bypass, and local rate hikes.

This chapter first examines the historical evolution of separations and traces how revenue requirements have been shifted to interstate toll. The next section discusses the economic policy issues that have resulted from the historical evolution of separations. The crucial issue is that the recovery of NTS costs through usage-sensitive rates is not sustainable in a competitive environment. The final section of the chapter offers conclusions and recommendations for policy.

Acknowledgments: This chapter has benefited from the comments of numerous people within ATT Communications; however, it expresses the author's views and should not be construed to reflect those of ATT Communications. Additional comments were provided by James Bradley, Sherrie Rhine, Ronald Wilder, and the Applied Microeconomics Workshop at the University of South Carolina. Special thanks are also extended to Edward P. Kittinger, who provided valuable information.

Historical Evolution

Background

In very broad and perhaps simplistic terms, separations may be considered a problem in the allocation of joint and common costs.[3] In general, it is impossible to assign joint and common costs to one product line based solely on marginal cost-pricing principles. Instead, some method of allocating these costs between services must be devised. It should be noted that there is no single economically correct method for allocating these costs; therefore, any method is arbitrary.

Compounding the problems inherent in joint and common cost allocation is the fact that separations occurs not for economic considerations but to fulfill the requirements of state and federal regulators. From a technological perspective, there is no difference between intrastate and interstate long distance, but an allocation must still be made because of regulatory considerations. Because of these inherent difficulties, this problem has been described as an "economist's nightmare for large volumes of true overhead and joint cost have to be sliced into economically dubious portions."[4]

Every industry faces the problems of joint and common costs allocation, but the implications are perhaps more profound in the telephone industry, since the vast majority of total costs cannot be readily assigned to any jurisdiction on a cost-causative basis. However, it is possible to assign them to customers on a cost-causative basis (see note 2). Therefore, any change in separations has policy implications because it shifts revenue requirements.

A joint board composed of three FCC commissioners and four state regulatory commissioners is responsible for providing guidelines, subject to full FCC approval, for the allocation or the separation of these costs between jurisdictions. These guidelines, which are considered a bible for those who must perform separations studies, are contained in Part 67 of the FCC rules and are commonly known as the *Separations Manual*. In essence, the *Separations Manual* lays out guidelines for allocating costs. If appropriate, a cost is assigned wholly to state or interstate. When this is not possible, the *Manual* provides a mathematical formula for allocating costs.

Separations Procedures and Changes in Cost Allocation

The need for allocating costs between intrastate and interstate jurisdictions for rate-making purposes was first recognized in 1913 in the Minnesota rate cases. Although these rate cases did not deal with the then-infant telephone industry, the precedent established was used in the development of the separations process.[5]

Early in the history of the telephone industry, long-distance service was considered an add-on to local service, and the only items used to compute the cost of long-distance service were the toll switching equipment trunks and lines. All of the local plant costs, including NTS costs, were considered to be part of local-exchange operations; hence, the costs were allocated to intrastate. This method was known as the board-to-board theory of separations (see Figure 13–1).

The board-to-board theory was the guiding principle in separations until 1930. In the landmark case, *Smith v. Illinois Bell Telephone Company,* the Supreme Court declared the board-to-board theory invalid. The Court ruled against this method because it ignores the fact that certain local equipment is also used to provide interstate toll service. However, the Court recognized the problems associated with joint and common cost allocation and stated:

> While the difficulty in making an exact apportionment of the property is apparent and extreme nicety is not required, only reasonable matters being essential, it is quite another matter to ignore altogether the actual use to which the property is put.[6]

As a result of this decision, the board-to-board theory was replaced by the station-to-station concept. Under the station-to-station concept, the local exchange is viewed as an integral part of the long-distance network. Therefore, some costs previously deemed wholly exchange must now be allocated in part to interstate with the remainder to intrastate (see Figure 13–2).

The *Smith v. Illinois Bell* decision, as applied, led to the allocation of costs by some measure of use, although this was not explicitly required.

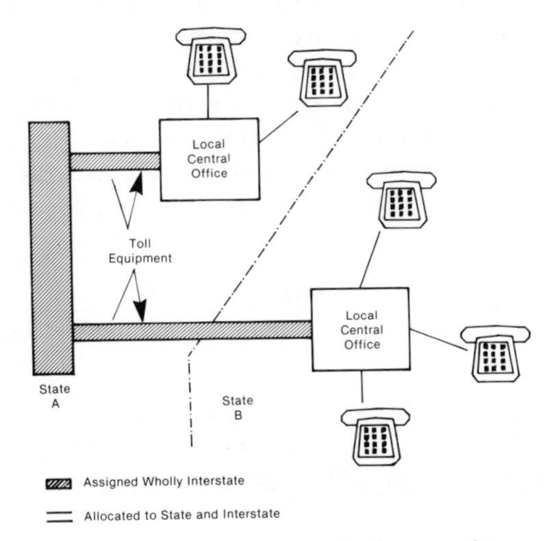

Figure 13–1 The Board-to Board Theory of Separations

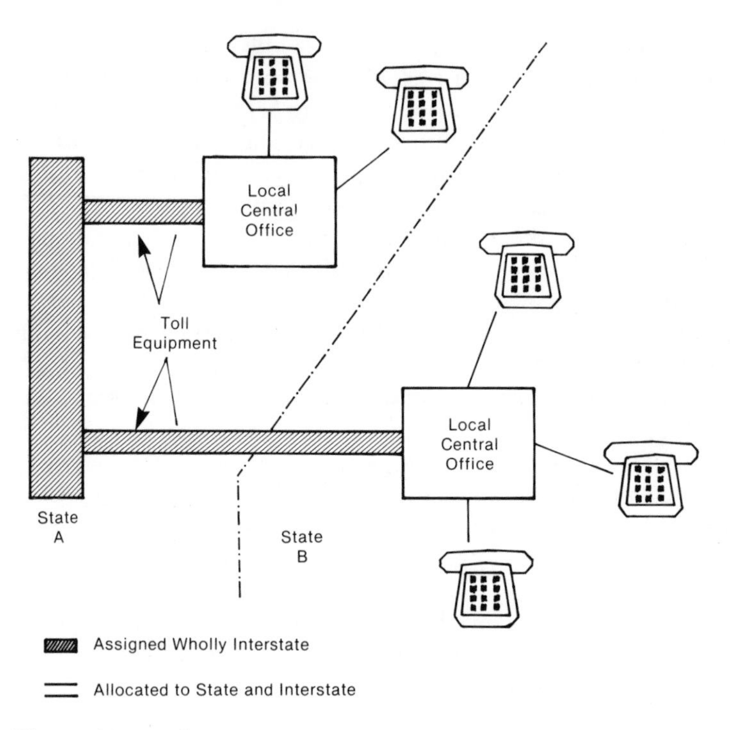

Figure 13–2 The Station-to-Station Theory of Separations

However, the term *use* was not strictly defined and has been periodically reinterpreted. A variety of use measures have been employed, ranging from simple relative use to a complex weighted use formula.

A good example of how the definition of use has been altered is the method by which non–traffic-sensitive (NTS) costs have been allocated. When the first *Separations Manual* was published in 1947, subscriber line use (SLU) was considered the proper definition of use. The SLU for interstate use is

$$\text{SLU} = \frac{\text{interstate toll minutes}}{\text{interstate toll minutes} + \text{state toll minutes} + \text{local minutes}}$$

It should be noted that the SLU allocator is just a simple relative-use measure.

As a result of the Ozark Plan, adopted in 1971, NTS costs were allocated by the subscriber plant factor (SPF). The SPF is calculated as follows:

$$\text{SPF} = \text{SLU} \times (2\text{CSR} + .85)$$

where SLU = subscriber line use;
CSR = composite station ratio.

The justification for the CSR was initially stated as the suppression of toll use relative to local use, because the toll price structure required a correction that measured use as if toll service were provided on a flat-rate basis. The CSR was supposed to provide that correction.[7] This justification appears rather strained, and the Federal Communications Commission has recently stated:

> SPF was developed as part of a Federal-State Joint Board negotiated agreement [that was] result oriented. SPF was designed to provide a specific level of compensation to local telephone companies for interstate use of facilities which are also used for local exchange service. As such, SPF embodies a heavy weighting of interstate message service.[8]

The impact of changing from the SLU to the SPF method of allocation is dramatic. From the same data, an SLU of 4.9 percent and an SPF of 16.29 percent for interstate toll are obtained.[9] On average, the effect of SPF is to increase the interstate allocation by a factor of about 3.3, as contrasted to a relative-use measure such as SLU.

Over the years, the method for allocation of traffic-sensitive costs, such as switching equipment, has also been altered. The effect of this change was also to allocate more costs to interstate. However, this change was not so dramatic as the case of NTS costs.

Separations procedures have been altered, over the years, in the face of changing political and economic realities. As new plans were adopted, a variety of political considerations were also mixed with the economic ones. During this time, the cost of providing long-distance service fell compared to the cost of local service because of technological innovation and the realization of economies of scale. Therefore, to keep local rates low, increasing amounts of cost were allocated to interstate long-distance service.[10]

Results and Implications of Separations Changes

Critics of separations have argued that one reason for these separations changes is that the technological requirements of long-distance service are different from those of local service, so the plant had to be designed for the more demanding service, long distance. However, the magnitude of this is unclear, since the increased demand for local service as well as network engineering considerations for large cities also require a more sophisticated local-exchange network.[11]

The amount of NTS costs allocated to the interstate jurisdiction has increased dramatically over the years. This shift has occurred because of both increased demand for long distance (use) and changes in the *Separations Manual.* The years of the major separations changes that increased the weighting of relative use (SLU) are summarized in table 13–1. As noted in the table, the various separations changes have placed a greater weight on usage. The implication of this weighting is that the percentage of NTS costs allocated to interstate toll has far outpaced the growth in demand as measured by SLU. This relationship is illustrated in Figure 13–3.

In order to prevent additional loading, or allocations of NTS costs to interstate, the FCC ordered the percentage of NTS allocations frozen at 1981 levels. This is the so-called frozen SPF allocator. This additional loading would have resulted from increasing demand for interstate long distance. Recently, the FCC ordered that the amount of NTS costs allocated to interstate be set at 25 percent for each state after an eight-year transition period beginning January 1, 1986. Currently, SPF differs among companies by state, ranging from about 10 percent to 85 percent. These differences are based on the relative usage of interstate toll, plus the CSR differences.

Policy Issues

Theory

Since the separations process is used to allocate costs and to determine price, it is at the heart of a variety of complex and interrelated issues—economic efficiency, bypass, and universal service, to name a few. This section discusses these issues.

Rholfs estimated that because of the separations process, the price of

Table 13–1
Major Separations Changes

Separations Plan	*Weighting Factor*
1943–52: local access costs allocated based on minutes of use	1.0 times SLU
1952–65: Charleston Plan	1.8 times SLU
1965–69: Denver Plan	2.5 times SLU
1969–70: FCC Plan	3.2 times SLU
1971–82: Ozark Plan	3.3 times SLU
April 1982–present: San Francisco Plan	Frozen SPF (1981 value)

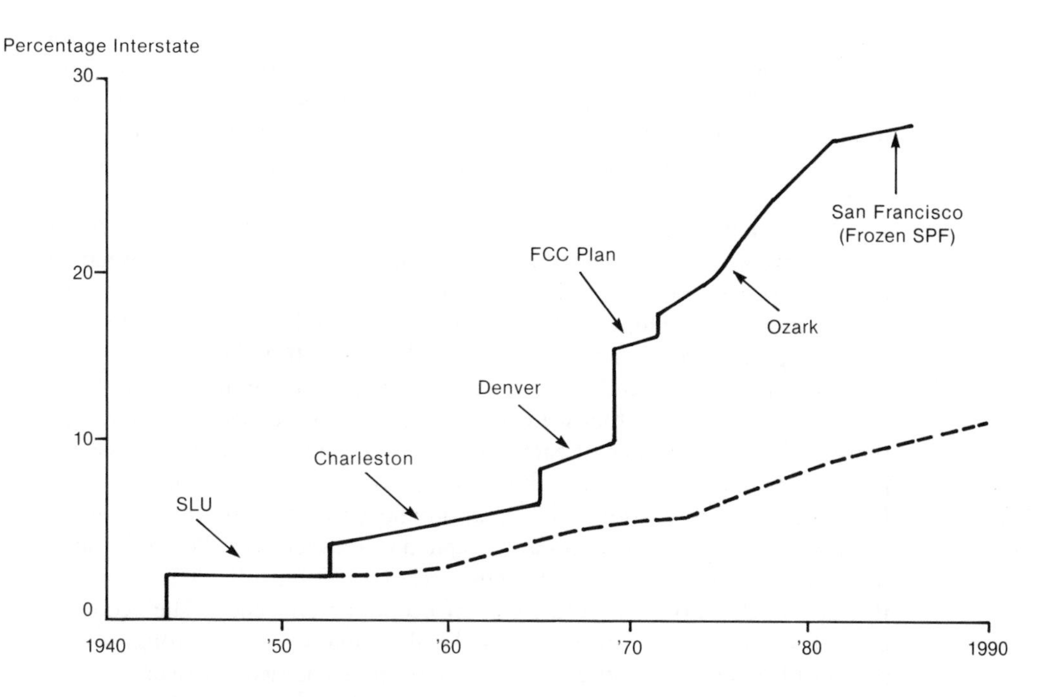

Figure 13–3 Non-Traffic Sensitive Plant Allocated to Interstate

interstate long distance is at least 50 percent greater than its marginal cost, and the price of local service is about 50 percent below its marginal cost. This would lead the casual observer to state that long distance subsidizes local service.[12]

However, the term *subsidy* is emotionally charged and should be used with caution because there are economic justifications for a departure from marginal cost. First, Ramsey prices (optimal departures from marginal cost) may be necessary to ensure the viability of the firm if economies of scale are present. Second, in the telephone industry, some degree of externalities may exist. That is, your subscribing to local-exchange telephone service in state A benefits me, a subscriber to local-exchange telephone service in state B, because I now have the ability to call you if I should so desire. Consideration of externalities suggests that it would be reasonable for me to bear some of the cost of your local-exchange service in the toll rates I pay when I call you. To date, however, no empirical evidence has been found that would justify the magnitude of the current subsidy.[13]

One approach to estimating the effects of these departures from marginal cost is James Griffen's 1982 study. Allowing for the foregoing qualifications, Griffen found that the welfare loss from the current pricing scheme was $1.8 billion a year for interstate long-distance service.[14] Wenders and Egan extended the work of Griffen to include intrastate toll markets. They found the welfare loss to be $15 billion to $20 billion. If the welfare effects for local-exchange pricing had been included, the loss would have been even higher.[15]

A different approach to the measurement of the economic losses found in the current telecommunications rate structure is contained in a Wharton Econometrics Associates study of 1983, which considered the macroeconomic effects of moving to efficient pricing. It found that real GNP over a five-year period would increase by $37 billion, to $46 billion.[16]

The current price and cost structure in the telecommunications industry introduces a distortion that leads to the welfare loss discussed earlier. Currently, a large percentage of each long-distance carrier's revenues is used to pay access charges to local telephone companies for the use of their facilities. For ATT Communications, this is about 60 percent of revenues. For the most part, these access charges are designed to recover NTS. Presently, some of these costs are recovered through a use charge on toll rates. However, this approach ignores the principles of economic efficiency. The economically efficient method to recover these local company costs is through a two-part tariff. Under such an approach, a flat fee is charged to cover the NTS costs, and traffic-sensitive costs are recovered on a usage basis.[17]

It should be noted that such a tariff structure is based on the principles of cost causation; that is, costs are recovered according to who caused them. Since NTS costs are related to hooking a customer up to the network and do not vary with use, it is appropriate that the customer or end user should pay those costs.

Application

The current pricing system ignores these economic principles, since a large part of NTS costs is recovered according to use. As a result, the large users carry the greatest burden of these NTS costs and may pay many times their own NTS costs. In fact, the small user is subsidized by the large one. A symptom of this distortion is the phenomenon known as bypass. Simply put, bypass occurs when alternatives to local-exchange carrier facilities are used to complete telephone calls. These alternatives include either the construction of private facilities or the lease of private lines from the local company. Recently, a variety of large users, such as banks, brokerage houses, and other such institutions, have elected to use bypass systems to achieve dramatic cost savings. It should be noted that these large toll users have the greatest incentive for bypass, because NTS costs are allocated according to toll use. Therefore, the larger the toll user, the larger the share of the burden it bears.

The result is that, at some point, it becomes more economical to make use of a bypass system than to use the local phone company.[18]

Bypass and the threat of bypass have attracted a great deal of attention from policymakers lately, because the use of toll service is highly skewed. For example, 10 percent of the business customers account for 80 percent of the revenues.[19] Therefore, the loss of a few large users (those with the incentive for bypass) will leave the local company with a revenue shortfall that necessitates rate increases.

This raises an important question: Is bypass economic or uneconomic? Bypass is considered economic if the cost of the bypass system is lower than that of the local telephone company or if the new system offers a capability not previously available. Bypass is said to be uneconomic if the bypass system is more costly than the local telephone company's network but is cheaper for the user because of pricing distortions. In this case, the bypass network is used by a large toll user group to avoid carrying a heavy burden of the non–traffic-sensitive costs. This represents a waste of resources, since unnecessary duplication occurs.

Proposed Solutions

Because uneconomic bypass occurs when prices are not aligned with marginal costs, the appropriate policy tool to discourage this type of bypass is the two-part tariff discussed earlier. Under this system, NTS costs are recovered by a flat charge assessed on each end user, and traffic-sensitive costs are recovered according to use. Since each customer is now paying only its own NTS cost, there is no cross-subsidization and no incentive for uneconomic bypass.

Recently, consumer groups and others have charged that correcting the rate structure distortions would threaten universal telephone services by increasing local rates. However, this charge has not been supported by empirical evidence. Perl found that the elasticity of demand for residential access to the local network is −.06. He also found that if measured service is available, the demand becomes even more inelastic,[20] and that the elasticity of demand differs across income groups and other demographic factors. Among the groups most affected are those with low incomes. One alternative offered to cushion the blow is targeted subsidies aimed at those who are harmed by increased local rates. Under such an approach, universal service may be maintained as the subsidy from toll to local is removed.[21]

In Docket 78-72, the FCC has attempted to move toward the two-part tariff approach discussed earlier. Currently, multiline business customers pay a maximum of $6 per month per line in end-user charges to recover a portion of NTS cost that was allocated to the interstate jurisdiction by separations.

Single-line business and residential customers pay $1 per month in end-user charges as of June 1985 and $2 per month as of June 1986.

Because of the uproar over end-user charges, the FCC scaled back the end-user charges for single-line business and residential customers from its original plan. Most states have not elected to use end-user charges to recover those intrastate NTS costs not recovered through exchange rates.

Summary

The purpose of this chapter has been to provide the reader with an overview of the jurisdictional separations problem and its relationship to economic efficiency. Besides providing an introduction to the nature of the problem, the chapter has reviewed the historical development of separations and has introduced the economic and public policy issues raised by past separations policies.

The basic economic issues raised are twofold. First, the amount of non–traffic-sensitive costs allocated to interstate toll has been considerably greater than the marginal costs. Second, and currently the most pressing issue, the method by which a large portion of non–traffic-sensitive costs are recovered is not economically correct, since these costs are recovered by a tax on toll use rather than by a flat fee paid by all telephone subscribers. The symptom of this departure from economic efficiency is uneconomic bypass. An effort to correct the problem, at least on the federal side, is seen in recent FCC orders mandating end-user charges.

It is hoped that this discussion will give the reader a better appreciation of the relationship between separations and telecommunications pricing issues. It may be safely said that an understanding of the separations process and its relationship to economic pricing is the Rosetta Stone for this complex and important area.

Notes

1. Non–traffic-sensitive costs are those costs that are incurred to connect a subscriber to the telephone network. These costs, which are found in the local telephone company plant, include the wire connecting the customer to the central office (the local loop) plus a part of the switching equipment. The difference between NTS and traffic-sensitive (TS) costs (those costs that vary with use) is crucial to current telecommunications pricing issues.

2. Congressional Budget Office, *The Changing Telephone Industry: Access Charges, Universal Service and Local Rates* (Washington, D.C.: U.S. Government Printing Office, 1984), 7. The NTS costs allocated interstate are currently recovered

through both end-user charges and toll rates; however, the bulk is recovered through toll rates.

3. In general, a joint or common cost occurs when the same input is used to produce more than one output. The classic example is Alfred Marshall's wool and mutton, which are outputs from the single input, the sheep. In the case of the telephone industry, the common input is the local-exchange carrier. These facilities are used to provide local service and connections to the long-distance network.

Historically, the telephone industry has been viewed as providing two basic products: local and long-distance calls. However, local-exchange companies also provide a third product: customer access. This gives a subscriber the ability to both send and receive calls. If customer access, the ability to send and receive calls, is considered a product, then the amount of unassignable costs drops dramatically and reduces the need for the "art" of cost allocations. This line of reasoning is well articulated in the direct testimony of John T. Wenders, Florida Public Service Commission, Docket No. 820S37-TP, August 1, 1984, 39–41.

4. William G. Shepherd and Clair Wilcox, *Public Policies Toward Business* (Homewood, Ill.: Irwin, 1979), 304–5.

5. A history of separations may be found in Richard Gabel, *Development of Separations Principles in the Telephone Industry* (East Lansing, Mich.: Institute for Public Utilities, 1967). Gabel's analysis makes use of the capture theory of regulation. Another view is found in Anthony Oettinger and Carol Weinhaus, *The Traditional Players, Stakes and Politics of Regulated Competition in the Communications Infrastructure of the Information Industry* (Cambridge: Harvard University Press, 1981).

6. As quoted in Gabel, *Development of Separations,* 24.

7. Perhaps the most Byzantine part of separations is the composite station ratio. The CSR is computed as follows:

> (1) the nationwide, industrywide average interstate initial 3 minute station charge at the study area average interstate length of haul to (2) the nationwide, industrywide average total toll initial 3 minute station charge at the nationwide average length of haul for all toll traffic for the total telephone industry.

Quoted from the *Separations Manual* in Anthony G. Oettinger and Carol L. Weinhaus, *The Federal Side of Traditional Telecommunications Cost Allocations* (Cambridge: Harvard University Press, 1981), 40.

8. *In the matter of Offshore Telephone Company,* para. 20, n. 12, Docket No. 21396 (September 21, 1983).

9. These figures are found in Ibid., 47. This source also provides an analysis of the distributional effects of separations.

10. See Oettinger and Weinhaus, *Traditional Players.*

11. Alfred E. Kahn, "The Next Steps in Telecommunications Regulation and Research," *Public Utilities Fortnightly* (July 19, 1984): 14. This article provides an excellent discussion of areas in need of research in the telecommunications regulation arena.

12. Jeffrey Rholfs, "Economically Efficient Bell System Pricing," Bell Laboratories Discussion Paper, January 1979, 65–66. This figure may be somewhat lower

because of the imposition of end-user charges and the phasedown of customer premise equipment to an inside wire.

13. Kahn, "The Next Steps," 15.

14. James M. Griffen, "The Welfare Implications of Externalities and Price Elasticities for Telecommunications Services," *Review of Economics and Statistics,* Vol. 64 (February 1982): 65.

15. John T. Wenders and Bruce L. Egan, "Improving Economic Efficiency in State Telecommunications Markets," Paper presented at Rutgers University Advanced Workshop on Regulation and Public Utility Economics, August 1984, 4.

16. Wharton Econometric Forecasting Associates, "Impact of the FCC Access Charge Plan on the U.S. Economy" (Philadelphia: Wharton Econometric Forecasting Associates, November 1983).

17. Currently, there is a vast literature on multipart tariffs. The seminal article is Walter Oi, "A Disneyland Dilemma: Two-Part Tariffs for a Mickey Mouse Monopoly," *Quarterly Journal of Economics* 85 (Feb. 1971) 77–96.

18. The two most comprehensive bypass studies to date are Gerald Brock, "Bypass of the Local Exchange: A Quantitative Assessment," FCC Office of Policy and Plans Working Paper, September 1984, 11–14; and BellCore, "The Impact of End User Charges on Bypass and Local Telephone Service" (Morristown, N.J.: Bell Communications Research, September 1984). Brock developed a model that shows that no equilibrium price can be obtained unless a flat end-user charge is in place. The BellCore study discusses the welfare effects of end-user charges.

19. John R. Meyer Wilson, R., Baughcum, A., Burton, E., and Caouette, L., *The Economics of Competition in the Telecommunications Industry* (Cambridge, Mass.: Oelgeschlager, Gunn and Hain, 1980), 186.

20. Lewis J. Perl, "Residential Demand for Telephone Service, 1983" (White Plains, N.Y.: National Bureau of Economic Research, 1983), 8.

21. Ibid. 28.

References

BellCore. "The Impact of End User Charges on Bypass and Local Telephone Service. Morristown, N.J.: Bell Communications Research, September 1984.

Bolter, Walter G. "The FCC's Selection of a 'Proper' Costing Standard After Fifteen Years—What Can We Learn from Docket 18128?" In Harry Trebling, ed., *Assessing New Pricing Concepts in Public Utilities*. East Lansing: Michigan State University, 1978.

Brock, Gerald. "Bypass of the Local Exchange, A Quantitative Assessment." FCC Office of Policy and Plans Working Paper, September 1984.

Congressional Budget Office. *The Changing Telephone Industry: Access Charges, Universal Service and Local Rates*. Washington, D.C.: U.S. Government Printing Office, 1984.

Cornell, Nina A.; Pelcovits, Michael H.; and Brenner, Steve. "A Legacy of Regulatory Failure." *Regulation,* Vol. 6 (July-August 1983): 37–42.

Fenton, Chester C., and Stone, Robert F. "Cost Allocation and Rate Structure: Concepts and Misconceptions." *Public Utilities Fortnightly,* Vol. 106 (July 3, 1980): 15–22.

Gabel, Richard. *Development of Separations Principles in the Telephone Industry.* East Lansing, Mich.: Institute for Public Utilities, 1967.

General Accounting Office. *Developing a Domestic Common Communications Policy: What Are the Issues?* Washington, D.C.: U.S. Government Printing Office, 1979.

———. *Legislative and Regulatory Actions Needed to Deal with a Changing Domestic Telecommunications Industry.* Washington, D.C.: U.S. Government Printing Office, 1981.

Grace, Martin F. "Access and the Demise of Separations and Settlements." *Public Utilities Fortnightly* (September 1, 1983) 17–22.

Griffen, James M. "The Welfare Implications of Externalities and Price Elasticities for Telecommunications Service." *Review of Economics and Statistics* (February 1982):59–66.

In the matter of Offshore Telephone Company, Docket No. 21396, n. 12, September 21, 1983, para. 20.

Johnson, Leland L., *Competition and Cross-Subsidization in the Telephone Industry.* Santa Monica, Calif.: Rand Corporation, 1982.

Kahn, Alfred E. *The Economics of Regulation,* 2 vols. New York: Wiley, 1970, 1971.

———. "The Next Steps in Telecommunications Regulation and Research." *Public Utilities Fortnightly* (July 19, 1984): 13–18.

———. "The Road to More Intelligent Telephone Pricing." *Yale Journal on Regulation* 1, 1984 139–57.

Katz, Michael L., and Willig, Robert W. "The Case for Freeing AT&T." *Regulation,* Vol. 6 (July-August, 1983): 43–49.

Majority Staff of the Subcommittee on Telecommunications, Consumer Protection, and Finance of the Committee on Energy and Commerce U.S. House of Representatives. *Telecommunications in Transition: The Status of Competition in the Telecommunications Industry.* Washington, D.C.: U.S. Government Printing Office, 1981.

Meyer, John R., Wilson, R., Baughum, A., Burton, E., and Caouette, L., *The Economics of Competition in the Telecommunications Industry.* Cambridge, Mass.: Oelgeschlager, Gunn and Hain, 1980.

Oettinger, Anthony G., and Weinhaus, Carol L. *The Federal Side of Traditional Telecommunication Cost Allocations.* Cambridge: Harvard University Press, 1981.

———. *The Traditional Players, Stakes and Politics of Regulated Competition in the Communications Infrastructure of the Information Industry.* Cambridge: Harvard University Press, 1981.

Oi, Walter. "A Disneyland Dilemma: Two-Part Tariffs for a Mickey Mouse Monopoly." *Quarterly Journal of Economics* 85 77–96.

Ordover, Janusz, and Willig, Robert W. "Pricing of Interexchange Access: Some Thoughts on the Third Report and Order in Docket No. 78-72." Princeton University Working Paper, June 1983.

Perl, Lewis J. "Residential Demand for Telephone Service, 1983." White Plains, N.Y.: National Bureau of Economic Research, 1983.

Rohlfs, Jeffrey. "Economically Efficient Bell System Pricing." Bell Laboratories Discussion Paper, January 1979.

Separations Manual. Washington, D.C.: National Association of Regulatory Utility Commissioners, 1971.

Shepard, William G., and Wilcox, Clair. *Public Policies Toward Business.* Homewood, Ill.: Irwin, 1975.

Wenders, John T., and Egan, Bruce L. "Improving Economic Efficiency in State Telecommunications Markets." Paper presented at Rutgers University Advanced Workshop on Regulation and Public Utility Economics, August 1984.

———. Direct testimony before the Florida Public Service Commission, Docket No. 820S37-TP, August 1, 1984, pp. 39–41.

Wharton Econometric Forecasting Associates. "Impact of the FCC Access Charge Plan on the U.S. Economy." Philadelphia: Wharton Econometric Forecasting Associates, November 1983.

14
Cost Support for Market Pricing: Challenges to Local-Exchange Companies

Frank J. Alessio
Criterion Incorporated

There is no doubt that over the past decade, technology, expanding customer desires, and competition have reshaped telecommunications markets. Market incentives are now the driving force in telecommunications. Competitive entry is transcending all regulatory and jurisdictional boundaries. Most telecommunications markets (particularly those that generate most of the revenues) are contestable, increasingly rivalrous, and workably competitive.

Competition is here now and will expand in the future. Bypass is occurring now and will expand in the future. No telephone company (not even a franchised local-exchange company) can act as if it were insulated from competitive market pressures. Consequently, commissions can no longer regulate any telephone company, telecommunication market, or rate structure as if it were insulated from competition.

Pricing Challenges

Many existing rate structures and price levels for intra-LATA telecommunications services are not viable for today's and tomorrow's market environment. New pricing initiatives and objectives used to be set by local-exchange companies (LECs) and public utility commissions (PUCs). Now, LECs will have to switch from rate-based pricing to market-based pricing. Future price structures and price levels will have to be deaveraged, unbundled, and segmented whenever necessary to gain market opportunities, respond to competitive vulnerabilities, and deal with particular regulatory problems. For example, there will have to be improved targeting of pricing objectives for access, which should reflect different market characteristics, if any, in different market segments—for example, wholesale (carrier) versus retail (end user) segments, public-switched versus private facilities, business versus residence segments, high-volume versus low-volume customers for access, and so forth. The pricing objectives for access should also distinguish between short-term expediencies and long-term objectives (for example, access charges based on jurisdictional separations procedures versus access charges based on market realities) and should include a strategy for moving from what is necessary and expedient in the short term to the longer-term objectives. There will also have to be improved targeting of pricing objectives for usage, which should reflect different market characteristics, if any, in different market segments—for example, high-volume versus low-volume customers for usage, voice versus data segments, distance segments, and so forth.

The PUCs must prepare for these changes also. They should pursue regulatory oversight in a manner realistically consistent with the dynamic conditions in the telecommunications marketplace. They should balance the legitimate needs of customers with those of interexchange carriers and LECs in the new environment. More precisely, they should modify current regulatory objectives and practices to reflect increasingly available competitive alternatives, rapidly changing technology, and expanding applications of telephone facilities to nontraditional telecommunications uses. The PUCs should be urged to allow more pricing flexibility than is currently possible under traditional rate-of-return and rate-base regulation. Maintaining a viable rate structure for the future will require that customer-specific contract pricing arrangements be allowed and that most services eventually be deregulated.

Costing Challenges

There will have to be modifications in the analytical tools used by LECs and accepted by PUCs—primarily cost analysis—to support market-based pricing. Five major points relate to perspectives on cost analysis by LECs in the new environment:

1. *Prices* and *costs* are not synonymous. The only role of costs in market-based pricing decisions is as a pricing floor—not the best price, not the maximum price, but the lowest acceptable price. Determining the cost of providing a service is never sufficient for determining the market price. In addition to cost information, a price is determined by market conditions—including value to the customer and current and potential competition. In fact, to avoid the easy temptation to employ cost-driven pricing decisions (for example, cost-plus approaches), it may be best to first identify a market price (independent of cost information) and then compare it to the cost information to determine whether the product or service can be offered profitably.

Thus, the first challenge is that managers and regulators should tip the balance away from a solely rate-base or revenue-requirement focus to more market-oriented cost analysis.

2. Although price and cost are not identical, they have the same conceptual foundation: both are value concepts. Price is a measure of value received by the customer. Cost is a measure of value forgone by the supplier—value forgone by using resources to produce a particular product or service. Cost is the opportunity value of the resources.
   The second challenge is that in a market environment, managers and regulators should recognize that the relevant cost information for pricing decisions is opportunity cost, not accounting costs and not revenue requirement calculations.
3. Opportunity costs are not book costs. Book costs reflect value at the time of acquisition, whereas opportunity costs reflect value at the present time or in some future time period. Book costs tend to be static, while opportunity costs tend to be dynamic. Book costs and opportunity costs will be identical through time only by coincidence.
   The third challenge is that in a competitive environment, managers and regulators should recognize that book costs are irrelevant for pricing decisions.
4. Opportunity costs are never embedded (historical) costs and are never fully distributed costs. Embedded costs are not actual costs; they are not the costs the company actually incurs. Operating expenses, the cost of money, depreciation, and capital refinancing are not fixed at past levels; all change through time. Fully distributed costs always result in umbrella pricing or barrier pricing, neither or which is acceptable in a competitive environment.
   The fourth challenge is that in today's competitive environment, managers and regulators should avoid using embedded costs and fully distributed costs as standards for pricing decisions.
5. Opportunity costs are most frequently referred to as incremental costs or marginal costs or avoidable costs.
   The fifth challenge is that in today's telecommunications market, managers and regulators should recognize and accept incremental costs, marginal costs, and avoidable costs as the relevant costs for pricing decisions by LECs as well as for rate oversight by PUCs.

Conclusion

None of the points and challenges just mentioned are likely to be news to many managers or regulators. Nevertheless, they are still the fundamental challenges underlying cost support for pricing in today's telecommunications market. The fundamental challenges are not in developing new techniques of cost analysis, as hard as the work will be; instead, the fundamental challenges are in adopting a market-based perspective: an application of market-based cost analysis by LECs, only now making the internal transition to market pricing, and an acceptance of market-based rules by regulators.

15
Post-Divestiture Telecommunications Pricing: Doing Business in the Competitive, Divested World

Jack Huber
BellSouth Services

The Business Challenge

The pre-divestiture, unified Bell System was relatively immune to the outside world. This is no longer the case. The new challenge in the telecommunications business is to anticipate and stimulate change in the outside world, change as it changes, and maintain profitability in the process. My point of view will be that of a local-exchange carrier.

The LEC business divides into traditional services and information transport services, but both use a common network. The business challenge for a local-exchange carrier, then, is to manage the traditional, slowly changing (and now by far the largest) part of the business and, through comon capabilities, to manage the rapidly changing and growing part—information transport services. In other words, an LEC must, at the same time, manage, using common technological and administrative resources, a mature, traditional business and a new, emerging business.

To get a feeling for the adjustments necessary to meet this challenge, it will be useful to discuss two different worlds—the averaged world and the differentiated world.

The Averaged World

The averaged and subsidized world, which began to end around the time of divestiture of the Bell System, had the following characteristics: Emphasis was on universal service. Customers were averaged. ("Most customers prefer . . ."), and pricing was averaged and subsidized by the formula: universe cost plus rate-of-return minus the long-distance subsidy divided by the universe size equals the price. There was also an adjustment of the average based on the value of services (businesses paid more than residences for equivalent services), and local service was subsidized by long distance, residence service by business service, and rural by urban.

The costs of research and development (R&D) were averaged, too. The aim was to maximize the nationwide revenue regardless of the relative contributions of local and long distance. Not much attention was paid to special customer needs, especially in data transport. Since there was a universal service emphasis, with almost no competition, the R&D was directed to the largest gains for the greatest number of customers. The Bell System was a vertically integrated monopoly in which regulation was nearly total, competition was extremely limited, and both services and responsibility were end-to-end. In this world, a system approach was applied, and R&D, product and service design, manufacturing, and services were all focused together. It was a cohesive industry in which economies of scale and scope allowed production and service efficiencies, which resulted in the most rugged, reliable, and efficient telephone system in the world. The focus was on traditional, telephone-oriented services, and the industry was generally responsive, but only narrowly so, to the demands of the information age.

The Differentiated World

At a time corresponding roughly with divestiture, the telephone business began to differentiate in several structural ways. This process is continuing, and divestiture caused some step changes. The differentiated world of the LECs now has the characteristics of a changing traditional business. The emphasis on universal service continues through regulation with averaging and subsidies. Exclusive franchises make the business appear to be the same, but the structure is eroding. Local regulators are beginning to follow the federal lead established in the long-distance arena. In Florida, for example, LATAs are subdivided to allow intra-LATA toll competition.

There is a trend toward cost recovery pricing. The ultimate goal is deregulation, but political realities guarantee a slow process. Resale and sharing of local service is allowed, both authorized and unauthorized. The R&D direction is also different. Individual RBOCs have different strategies and technology is controlled by equipment suppliers without a system approach.

In the vertical disintegration that was mandated, LECs got local transport and AT&T got long distance, customer premise equipment (CPE), and network equipment manufacturing. A multitude of new players now have entered various parts of the business, and the systems approach has been replaced by piece-parts with interface specifications.

Competition is encouraged in today's world—and services and responsibilities are fragmented. LEC equipment suppliers are also LEC competitors.

AT&T and Northern Telecom PBXs compete directly with the switching equipment that those companies supply to the LECs. Pre-divestiture economies of scale are eroding for the LECs because of LEC strategy differentiation and fragmented buying shares resulting in a lack of buying clout. It is a fragmented industry in which the focus is on both new and traditional services. Responsiveness to customer demands increases as the market differentiates. When fragmentation is accommodated with interface standards, overall costs increase, reliability decreases, and attention to special customer needs increases.

This two-world comparison is background. The business challenge for a local-exchange carrier in the present, increasingly differentiated world is to manage the traditional part of the business and, through common capabilities, the new part as well. The remainder of this discussion will focus first on the challenges presented by each of the two parts of the LEC business and then on the relationships between them.

Traditional Services

The traditional business is a challenge in itself because it is still, by far, the largest part of the business, producing most of the revenue. It is a mature market, yet it has potential. It is both protected and endangered by regulation, and competition continues to increase. The relationship between regulation and competition is one major area of the management challenge in the traditional services; two others are the maturity of the market and the market potential presented by the information age.

Regulation versus Competition

The prime example here is the bypass issue, in which loss of revenue means that LECs must raise local rates or interexchange carrier (IC) access charges, or both, and increasing IC access charges means more bypass and therefore more upward pressure on local rates. Regulators, with universal service goals, and LECs, with business market share desires, are both caught in the crunch.

The simplistic solution to bypass would be an estimated 140 percent increase in residence rates. Since this solution is not politically feasible, there is a mix of pricing adjustments and a stepped progression of subsidy reductions. Costs will very slowly shift to the user as subsidies are reduced by regulators. Access and usage have now been unbundled, and usage-sensitive pricing is on the increase. Deaveraging, though not prominent yet, is being discussed. With deaveraging, access charges would be a function of a customer's distance from the serving telephone office.

Solutions to these problems begin to look like market-price tactics, but regulators must be convinced of the wisdom of each small change. Shared-tenant arrangements, which reduce LEC revenue, receive little regulatory attention. As market maturity increases slow growth, experienced buyers, and a declining rate of innovations (at least at the present time) are the challenge. The market is the predominant revenue source with a high potential, and careful cost control is a standard strategy. In the market of the future, customer needs orientation is a necessary competitive response, with bypass, PBXs, local area networks (LANs), and shared-tenant arrangements focusing on customers. Market research is an essential technique in three areas:

Customer needs—both special and marginal needs must be considered.

Technology—network and CPE intelligence process is required because of loss of end-to-end control.

Competition—both reactive and cooperative opportunities are presented.

New products and applications must result from market research efforts, and plant modernization must maintain the state of the art, because the competition will continuously advance it.

Information Transport Services

This is the more dynamic part of the business, requiring the greatest flexibility and responsiveness and presenting opportunities that are not without problems left over from the averaged world. The management challenges are in research, product development, new service implementation, and the velocity of change. It is a question of transition management—from the services of today to the more capable services of tomorrow.

As in the traditional services, research is required on customers, technology, and competition; but all three areas are more differentiated. In product development, the regulatory time lag retained from the averaged world slows the response to market needs for new services. LEC; equipment suppliers are also competitors—they are not dependent on one LEC, they can provide features on their competing equipment first; and they face a differentiated LEC market.

Traditional economies of scale are more difficult to achieve when technology must be deployed in a differentiated market. Among the considerations are geographic clustering, customer segmentation and the revenue priorites this implies, and a necessary synergy with traditional services. Interfaces and standards have become external considerations because of the fragmented industry structure. Regulators are addressing interfaces and their implications; for example, protocol conversion is a public issue. Standards are affected by the widening equipment-supplier base and its wide customer base; off-network developments in LAN and PBX; and international entities such as the CCITT.

In the implementation of the new telecommunications age, quality is more difficult to maintain in an increasingly diverse product line with many small-scale products, training and administration costs increase, and enrichments must keep pace with competitor's state of the art. The velocity of change requires increased attention to customer education. Transition services on the existing network must be developed to seed the market for the future, and standard interfaces must be emphasized to minimize customer transition impacts.

The Linkage

The traditional and the information transport services are inextricably linked through hardware, administrative systems, technical expertise, the use of the network—and the fact that it is economic and accurate to transport and switch almost all kinds of signals, voice included, on a digital basis. The business challenge is to make the most economic use of this linkage in the service of differentiated customer needs.

The efficiency of this process is enhanced by the establishment of links in upstream planning—planning that is supported by market research as well as by the traditional methods of demand forecasting.

But the divested world is rapidly becoming more competitive. The allocation of resources thus becomes a function of competitive market dynamics. The bypass issue I mentioned earlier is already forcing new pricing structures in both business and residence services. This, then, is another LEC challenge—to develop and deploy competitive market techniques, in the information transport part of the business, certainly, but also in the traditional part. Segmenting the market, unbundling and deaveraging prices, and using willingness-to-pay measures will all play a part.

Finally, there is the linkage at the customer level. Customers must be made aware that new service options are available and that these options can improve the customer's competitive position. This linkage will serve the LECs by migrating customers from the restrictive environment of traditional services to the freer environment defined by information transport services. It will also serve customers better by meeting their special needs.

Recurring Themes

This chapter on doing business in the divested, competitive world has touched on several themes, many of them more than once, seem to encapsulate the current LEC business environment and its challenges:

Market orientation is most important in a competitive situation.

It is necessary to manage the synergy of traditional capabilities and new services.

Market research serves as the five senses of the business.

Customer education is essential for market development.

Continuing regulation is a mixed blessing.

Pricing evolution is essential if the trend to deregulation is real.

Nonstandards are what the customer pays for competition.

The LEC-supplier relationship is unsound in a competitive environment.

16
Pricing for Today's Environment

Clinton Perkins, Jr.
Southern Bell

There are probably as many differences of opinion on what pricing should be as there are people with opinions. One says: "I want the lowest price I can get." Another says: "I want the lowest price I can get and still get quality." Yet another says: "I'm not concerned with the price, I want quality." To be successful, telephone companies must pick the price level that meets all of our customers' needs and, at the same time, covers our cost of providing the service, produces a profit so that we can meet our investor needs, and meets the test of regulations. If you have ever walked a tightrope with one foot tied behind you, you know the challenge. In short, the move from a monopoly to a quasi-monopoly, quasi-competitive pricing environment will test the telephone companies' ability to catch on quickly to the sensitivity of pricing.

To meet these challenges, we must greatly expand our intelligence-gathering efforts, both within the companies and in the marketplace. BellSouth Services is setting up a market research/strategic structure design group to serve the two telephone companies (South Central Bell and Southern Bell). Our goal is to find out and know as much as we possibly can about the marketplace's telecommunication requirements as soon as we can.

To be responsive and produce the desired results—a healthy revenue stream for South Central Bell and Southern Bell—it is necessary to have more dedicated people in the pricing and cost functions. We are going to have to develop and be ready to use mechanization and computer programs to achieve the shortest possible response time. You will note that I have said *response time*. Please note, too, that I emphasize *proactive* instead of *reactive*. We must be out in front.

Pricers and costers must also be service-technical smart. They must understand what goes into making up and producing a service and what the accumulated cost is to produce that service. These pricers/costers must not be just acceptors. They must understand what they accept, be able to ask (and answer) questions about the information that is given them, and be able to make an overall constructive impact on the building of the service and the cost of producing it.

Setting Prices for Services

In setting the prices for the services that we, the telephone companies, provide, we must know what our costs are. Most important, we must know the price that the service will sell for in the marketplace. If we know the price and the cost, then we know whether or not the service will produce a profit or contribute to the financial viability of our enterprise. Price does not equal cost. Price produces a profit above cost to assure the financial viability of our enterprise.

Probably readers have their own terms for this idea: cost-plus pricing, target-profit pricing, going-rate pricing, or perceived-value pricing, to mention a few. I would like to be looked upon as a market pricer.

In my opinion, a full appreciation of consumer expectations is one of the most difficult aspects of pricing telecommunication services. Until recently, we were totally regulated. We were able to subsidize exchange service with revenues received from message toll, custom calling services, and so forth. Today that opportunity is rapidly being reduced. Therefore, we must expand our research and get better acquainted with customer expectations.

Factoring consumer price expectations into the price-setting equation could be helpful in increasing profits by helping the price setter discover his market. Customers may not be willing to pay higher prices, and lower prices might also reduce profit. Moreover, a company faced with raising prices because of costs or competition might phase in price increases while trying to adjust consumer expectations through better communications. Since negative customer reaction can raise costs, lower revenue, cause late payments and increase dunning costs for slow payers, and so forth, using of price expectations can help preserve or even increase profits.

In pricing telecommunication services, we must know what is expected, what is desired, and what is needed in the marketplace.

Further, we must recognize what Dr. Frank J. Alessio noted in a report entitled, "General Pricing Policy For Telecommunications Services":

> The principal pricing policy problem is that many existing rate structures and price levels for intraLATA telecommunications services are not viable for the future environment. To be made viable, future rate structures and price levels will have to be deaveraged, unbundled and segmented when possible to gain market opportunities, protect against competitive vulnerabilities and isolate particular regulatory problems. More specifically:
>
> 1. There will have to be less averaging in the rate structures and price levels for many services.
> 2. There will have to be improved targeting of pricing objectives for access. The pricing objectives for access should reflect different market characteristics, if any, in different market segments; e.g., wholesale versus retail (carrier versus end user), public-switched versus private facilities, busi-

ness versus residence, high volume versus low volume customers for access, etc. The pricing objectives for access should also distinguish between short-term expediencies and long-term objectives (e.g., access charges based on jurisdictional separations procedures versus access charges based on market realities.)

3. There will have to be improved targeting of pricing objectives for usage. The pricing objectives for usage should reflect different market characteristics, if any, in different market segments; e.g., high volume versus low volume customers for usage, voice versus data segments, distance segments, etc.

He further said:

Economically appropriate prices for all services must be based on market characteristics, which include the value of the service to customers, current and potential competition and the incremental cost of providing the service. Competitive and potentially competitive services should be priced to generate the largest long run contribution to the Company to maintain the financial viability of the Company and send appropriate price signals to the market. Non-competitive services should be priced to at least cover their incremental costs, except (if possible) the primary residence access line (and only one line per residence, and in a measured service environment (not a flat-rate line) and, perhaps, only to residence customers who cannot truly afford to pay a higher price). Thus for all services (again, except the primay residence access line as defined above), the price floor should be the incremental cost of providing service.

Determining Realistic Costs

We must know about the marketplace; but we must also know what our costs are. The price floor should be the incremental cost of providing service, because incremental cost is the economically appropriate cost for decision making. It is prospective decision making: it produces pricing decisions that include the planning period; it addresses service introduction or discontinuation decisions; and it addresses decisions on when and where to unbundle, deaverage, and segment rate structures. Incremental cost analysis requires information on the prospective costs associated with these decisions and reflects the costs that are a direct result of the decision being made (that is, it reflects the costs that are caused by the decision and does not include costs that are unchanged by the decision). Again, pricing decisions require incremental cost information because they are forward-looking decisions; thus, the cost information could reflect the costs that will be incurred to provide service in the future, when the prices will be in effect. Incremental costs reflect the prospective environment, including the future price levels for labor, capital, materials, and so forth. In addition, they include the methods and technologies of production that will be used in the future. Finally, incremental costs reflect the future value of the resources used to provide the service, which is, in turn, affected by the potential for their reuse.

To summarize, I believe that in the increasingly competitive environment, rate structures and price levels should be based on market characteristics, such as value to the customer and current and potential competition, and that they should be tested against the incremental cost of providing the service. Competitive and potentially competitive services should be priced to generate the largest long-run contribution. Embedded costs and fully distributed costs are not appropriate cost methods for pricing decision making. Incremental cost is the economically appropriate cost for decision making, because it reflects the costs that are caused by the decision.

Reference

Frank J. Alessio, "General Pricing Policy For Telecommunications Services," Report to BellSouth Services Inc., January 1984, pp. 1–2.

Part IV: Other Issues in Telecommunications

Robert Fortenberry emphasizes that divestiture and the move toward regulated competition have been driven by technological developments. Regulated competition was instrumental in achieving universal telephone service and will be the force behind achieving what he calls universal information service. However, regulation needs to give way to competition for the full benefits of technological change to be realized. Equal access is two-thirds complete among the Bell Companies in the four southeastern states. Prices must reflect costs, and where they do, the cost of long-distance and local service will be less than it has been in the cross-subsidized, partially regulated era of today.

In Chapter 17, the editors of this book discuss market power in the intrastate, inter-LATA toll market. For the most part the study is generic and methodological. A primary purpose of the Chapter is to specify the theory underlying the analysis of market power, including the theory of unregulated and regulated markets, both single and multiproduct. The focus is on precise definition of the concepts *market, market power, market shares,* and *natural monopoly*. The authors explain that market shares are relevant because they represent one, but only one, element of market power. The study also contains an analysis of how market shares are related to market power.

Terence Robinson develops an economic model to measure the value to society of selected technologies in their expected life span. Working within a framework that embraces the theory and concepts inherent in economic selection theory, revenue requirement regulation, the time value of money, and accounting theory, he moves from a depreciation to a capital recovery perspective that recognizes the heightened competition and explosive technology that now characterizes the telecommunications industry. Although his chapter employs significantly more advanced quantitative methods than the other chapters in this part, it is worth the reader's greater effort, because it deals with an important issue that surely will grow substantially in importance until it is resolved. Anyone who is interested in copper cable and fiber technology is encouraged to peruse this chapter.

The chapter by William Hederman illustrates the universality of the telecommunications experiences—especially the move toward increased competition—and the common problems among regulated companies and industries. Dr. Hederman argues that many of the characteristics of telecommunications markets have their counterparts in natural gas markets. In particular, he feels that the problems of local bypass, cross-subsidization, equal access (mandatory carriage), unbundling of services, and market domination in certain markets are common to both industries. He concludes with the observation that the experiences in telecommunications may provide a basis for analysis and solutions in natural gas markets as they become deregulated.

The chapter by Professors R. Carter Hill, Albert L. Danielsen, and David R. Kamerschen, assesses the feasibility of modeling the state-level economic impacts that would be associated with altering carrier access and customer line charges. Wharton Econometric Forecasting Associates has undertaken such an analysis at the national level. Although most economists correctly predicted the direction of the changes (that a more nearly cost-based system of telecommunications pricing would tend to promote a more efficient utilization of the nation's resources), many were surprised by the magnitude of these changes in the Wharton model. But what guidance this national study provides at the state level is not clear. Therefore, Hill, Danielsen, and Kamerschen did a survey of state macroeconomic modeling groups in nine southern states to see if the same phenomenon could be investigated at a more dissaggregated level than Wharton had attempted. Although the authors did not actually attempt to model these changes, they did provide a path through the jungle for future researchers to traverse in such an undertaking.

17
A Methodological Study of Market Power and Market Shares in Intrastate Inter-LATA Telecommnications

Albert L. Danielsen
David R. Kamerschen
University of Georgia

One of the most vital and challenging issues in this transitional era in the telecommunications field is that of assessing the market power of the numerous competitors.[1] Since the telecomunications industry is rapidly undergoing a transition from a largely regulated to a more deregulated environment, and since the market participants are being treated unequally, it is now imperative to assess accurately the actual and potential market power of the rivals.[2]

The assessment of market power in a regulated milieu is different than it is in an unregulated one because there is an implied immunity from the antitrust laws. While the degree of immunity is debatable, there is no question that historically there was pervasive regulation under a public interest standard.

Similarly, with deregulation it is particularly important to assess the degree of market power on a more disaggregated basis than the entire nation. This information is particularly needed at the statewide level, as state public service commissions must make crucial decisions that require knowledge of the market power of the competing firms. These commissions have a mandate to promote the public interest—which includes consumers as well as shareholders—and significant market power can substantially influence the public interest.

Study Design

Major Objectives of the Study

This study is generic and the orientation is primarily methodological. Operating within that framework, we are attempting to accomplish three major objectives. The first is to specify the theory that underlies the analysis of market power. This part of the study involves defining the terms market, market power, and market shares, and explaining precisely how market shares are related to market power. The second objective is to determine the feasibility of establishing a system to monitor the market shares of AT&T-Communications (hereinafter referred to as AT&T-C) and the other interexchange carriers (hereinafter referred to as OICs). And finally, if such a monitoring system is feasible, the third objective is to specify the sources of data and methods that could be used to establish a monitoring system.

Market shares are relevant because they represent a possible element of market power. In fact, a high market share is sometimes considered a necessary condition for a firm to be able to exercise market power, or to raise prices above what they would be under competition. However, there are several different measures of market shares, and both their levels and trends may be relevant in assessing the importance of market shares as indicators of market power. In addition, market shares are but one piece of that mosaic called market power, or, in other words, market shares are only one set of a host of factors that should be considered in a comprehensive assessment of market power.

Scope of Study

Although this study is primarily generic, the state of Georgia will be used in most discussions related to specific markets, institutions, regulations, and of the data available to assess market power and market shares. However, it is anticipated that the basic methodology could be extended to any state. In most cases the data required to assess each factor of intrastate market power quantitatively are not available, and so we outline the factors that should be examined in a qualitative assessment of market power. We also indicate what, if any, national data are available to assess each factor that may confer power upon a market participant.

Our focus may be contrasted with many previous studies that have focused on determining the extent of scale economies and cost elasticities with respect to output. We are not concerned with the ratio of marginal to average cost, but rather with the end results of those costs and the revenue streams that are currently being determined by a mixture of competition and regulation.

The Basics of Markets, Market Power, and Market Shares

Market power is a term used to indicate the absence of competition or the potential existence of monopoly elements in a market. Broadly, competition exists when buyers have alternative sources of supply, either actual or potential, at prices they regard as comparable. A pure monopoly exists if there is only one seller. There may be monopoly elements in a market when buyers have few alternative sources of supply.

The following is a general discussion of markets and a specific discussion of telecommunications markets. Formal and working definitions of market power and market shares are provided. Since market shares are often used as surrogates to measure the degree of market power held by the leading firm(s) in a market, there is also a discussion of the complex relations between market power and market shares.

Rigorous definitions of the terms market, market power, and market shares are required when making quantitative assessments of a market. Our task has been simplified considerably by the availability of an extensive body of literature on each of these concepts, developed mostly by economists, lawyers, jurists, and regulatory agencies. The definitions and concepts presented in this study draw heavily on the work of academic economists as well as that of scholars working in the private sector and in executive and judicial agencies such as the Federal Communications Commission (FCC), Department of Justice (DOJ), Antitrust Division, and the U.S. Supreme Court.

One purpose of this discussion is to explain how these entities or economic agents have interpreted the terms market, market power, and market shares. Our purpose is not to trace the epistimology of these terms, but we believe that current usages and interpretations are relevant to telecommunications markets. We draw heavily upon the antitrust literature and upon antitrust guidelines as presently administered by the DOJ. We believe this orientation is appropriate; as telecommunications markets continue on their present course toward a lesser degree of regulation, the antitrust implications will become ever more important. Telecommunications is in the process of moving from the tight direct regulation of both federal and state agencies to a looser regulation in which the indirect regulation of the antitrust laws is relevant. Clearly, as the regulatory authorities grapple with the problems of a less regulated industry, they will have to deal with the same conceptual issues regarding markets, market power, and market shares as the executive and judicial agencies have had to deal with in their surveillance of the less regulated industries.

Definition of a Market

General Definitions

Single and Multiproduct Markets. In a general sense, a market consists of buyers and sellers who are in close communication with one another, and transactions take place. A market may consist of buyers and sellers consuming and producing a single homogeneous (or fungible) product for which there are no close substitutes or more generally of sellers and buyers consuming and producing a relatively standardized product for which there are gaps in the chain of demand and supply substitutes compared to other markets.

While much of the received traditional economic theory and law is based on markets in which firms produce a single solo product, in reality most businesses produce many products. Most regulatory and antitrust issues involve these multiproduct enterprises. This multiplicity of products can take the form of a variety of physically dissimilar offerings, a wide variety of offerings of similar products (such as hats of different sizes) adapted to the desires of individual consumers, or just physically similar offerings sold at various places or times (see Evans, 1983, p. 132). Thus, in order to understand today's business world, definitions of multiproduct markets are absolutely essential. However, the economic principles of these markets are difficult to develop and comprehend, so there is less of a consensus on their proper treatment. Perhaps it is a reflection upon the economics profession, or simply an indication of how intractable the problems actually are, but even the principles of multiproduct markets are in a state of infancy.

Bases for the Definition of a Market. According to many economists and the U.S. Supreme Court, a market encompasses the primary demand and supply forces that influence a product's price and the geographic area that encompasses these buyers and sellers over some reasonable time frame. Similarly, according to the DOJ (one of the two principal federal agencies responsible for enforcing the antitrust laws and promoting competition, and, therefore, for assessing actual and potential market power.)

> . . . a market is defined as a product or group of products and a geographic area in which it is sold such that a hypothetical, profit-maximizing firm, not subject to price regulation, that was the only present and future seller of those products in that area would impose a 'small but significant and nontransitory' increase in price above prevailing or likely future levels. (DOJ "Guidelines", 1984, p. S-1)

It is also important to note that, subject to certain exceptions, the DOJ considers the smallest market in each case to be the relevant market.

Natural Monopoly. A so-called natural monopoly is of particular interest to this study inasmuch as it is *sine qua non* according to the public interest theory of regulation. Until recently, economists and legal scholars defined a natural monopoly as existing when scale economics are present. Baumol,

Panzar, and Willig (BP&W, 1982) have shown that economies of scale are not the exclusive determinants of whether a market is a natural monopoly. Rather, a given market's output is said to be an output-specific natural monopoly if that output can be produced less expensively by one firm than for any division of that output among two or more firms.

More generally, a market is a natural monopoly if for a market output level the product(s) can be produced most efficiently by one firm.[3] Put differently, a natural monopoly market exists if the cost-minimizing structure is a single firm (BP&W, 1982, p. 351) This means that economies of scale are neither necessary nor sufficient for a natural monopoly, suggesting that natural monopoly markets are significantly different and smaller than formerly believed. BP&W have shown that firms with sufficiently high fixed costs will have the necessary and sufficient (subadditive cost) conditions necessary for natural monopoly.

When Evans (1983) and Evans and Heckman (1984) estimated a multiproduct cost function for local and long-distance service for the Bell System from 1958–1977 they found that AT&T was not a natural monopolist (that is, it did not have a locally subadditive cost function).[4] However, all of the published estimates of which we are aware were done before divestiture and are therefore not necessarily applicable. It is instructive that the DOJ ("Comments", 1984, pp. 10, 39) believes that the interexchange telecommunications market is not a natural monopoly and should not be regulated in the long-run.

Factors to Consider in Defining a Market

Examination of the literature cited above clearly reveals that a market is multidimensional and has at least three important dimensions: *product, geography,* and *time.* While the product dimension tends to predominate in law and in economics, it is important to recognize that every market has both geographic and time dimensions as well. The geographical dimensions of a market are sometimes more important than the product dimensions. And in exceptional cases the time dimension may be the most relevant. Each of these are discussed in this section.

Product Dimension. Some important pronouncements from a U.S. Supreme Court case can be used to illustrate the present status and state of the art of determining the dimensions of a market. The Court has ruled that in order for two or more products to be regarded as part of the same market, the prospective buyers must regard the products as "reasonably interchangeable" (see DuPont, 1956, p. 394). The Court also specified that any evaluation of a market using this criterion must include an analysis of the price, quality, and the use of the product.

Use refers to a highly pragmatic way of identifying products to be included in a market (that is, by examining the uses of those products). *Quality* refers to the fact that reasonably interchangeable products should have similar characteristics. When examining the *price* variable, the Court uses the economist's concept of cross price elasticity of demand. The cross price elasticity of demand is defined as the percentage change in the quantity of one product purchased which is the result of a 1 percent change in the price of another product, assuming all other things are held constant.

Certainly the criterion of products being reasonably interchangeable is a good place to begin, but it is not an adequate basis for defining a market. The test of reasonable interchangeability emphasizes substitutability in demand or end use and is an appropriate criterion to use in defining the most narrow boundaries of a market. However, a correct definition of the relevant product market should also include close substitutes in production.

It should be emphasized that the concept of a "good" substitute, either in consumption or production, is meaningless without specifying the prices of the alternative products. For example, two products not considered reasonably interchangeable because of a large difference in their prices could become reasonably interchangeable if the price gap were to narrow. "Plain old telephone service" or enhanced digital communications may not be good substitutes for "plain old postal service" at present prices, but should their price gaps narrow they may become good substitutes.

Finally, the existence of numerous reasonably interchangeable products can actually demonstrate a lack of competition. When the prices of goods or services are priced close to their cost of production, there may be few viable substitutes. As the price increases, there are more and more acceptable alternatives, but these are "poor" alternatives which are only acceptable because of the large price differential, which was itself brought about by the exercise of market power. The alternatives can effectively compete only because of the high price charged for the superior alternative. This limitation is especially relevant in evaluating the use of cross price elasticities of demand as a tool for defining markets. Discovering that the demand for one service is very sensitive to the price of another may only mean that the price is already at a monopoly level.

A more relevant comparison could be obtained by examining the cross price elasticities at prices close to the costs of production for each of the services. In addition, a product market definition should consider costs of production and quality differentials. However, recognizing the difficulties involved in evaluating price differentials and using cross price elasticities, the courts have stressed the similarity of qualities and end uses in applying the "reasonably interchangeable" test.

Geographic Dimension. A market has geographic boundaries and may also be bounded by other forces that impede the flow of information and reduce the volume of transactions. On the buyers' side, the geographic market is limited to the area in which consumers can reasonably be expected to make purchases. That area can vary dramatically with the nature of the product or service. For sellers, the geographic market is bounded by the area in which they can compete. In many cases, transportation costs are the determining variable.

The importance of geographic boundaries is great for bulky and heavy products that have a low market value, such as ready-mixed concrete. Such space considerations increase the cost of products at the point of delivery, confer a competitive advantage on suppliers located near the end-user, and thereby limit the size of the market. Any other factor or force that lowers costs for some suppliers, or potential suppliers, but not for others, will affect the size and scope of the market.

Time Dimension. The time dimension is relevant because demand and supply elasticities generally increase over time, making competition more vigorous. When assessing the competitiveness of a market, the time period should be neither so short that it includes only the existing or incumbent firms, nor so long that it allows time for all potential changes in technology, demand, and tastes to occur. For if entry is relatively free, and firms are given enough time to adapt to new developments, almost any two firms could be regarded as potential competitors.

Definition of Telecommunications Markets

Definition of a LATA. A necessary predicate to understanding our approach to the appropriate definition of the statewide telecommunications market is to understand what is meant by the term LATA. On October 4, 1982, AT&T and the operating companies filed new service areas—known as LATAs (Local Access and Transport Areas). A LATA, generally the size of a large city and its surrounding rural areas, marks the geographic boundaries beyond which a Bell Operating Company (BOC) may not transmit telephone calls. The BOCs are authorized to (1) engage in both exchange and toll telecommunications, that is, transport local and toll traffic between telephones located within a LATA and (2) provide exchange access within a LATA (link a subscriber's telephone to the nearest transmission facility of an interexchange carrier). In other words, the seven Bell regional holding companies are currently permitted to provide only complete origination, transmission and termination of calls that do not cross a LATA. Calls that cross a LATA currently must be carried by an IC. Because some of these LATAs are quite large, a substantial amount of long-distance traffic occurs within LATAs. For instance, Nevada is a LATA; in all, twenty-four states have only one or two LATAs. On the other hand, Delaware is part of a LATA which includes Philadelphia, Pennsylvania.

The 161 LATAs proposed by AT&T followed the criteria contained in the Modification of Final Judgement (MFJ, approved by Judge Harold H. Greene of the federal district court in Washington D.C., on August 24, 1982) in that: (1) the areas serve common social, economic, and other purposes; (2) every point served by the divested operating companies shall be included within a LATA; (3) except with court approval, no LATA shall cross state lines.

Basic Criteria. Turning from the general to the specific, how should telecommunications markets be delineated? Such markets could be defined for each service or group of services according to purpose, and for alternative geographic areas in which the services are sold. But for the purpose of assessing the intrastate market power of a firm, the economic and legal realities of the 1980s need to be considered. Thus, the definition developed in this section is partly dictated by the present state of technological development in the industry, by federal-state regulatory jurisdictions, and by the availability of data on telecommunications markets. In addition, the relevant markets must be defined so that an assessment of the competitive situation is both meaningful and credible to an unbiased observer, and relevant to the problem at hand. Thus, the appropriate definition of a market depends upon the issue one is investigating. Since the charge of a regulatory commission is to promote the public interest, an appropriate definition of a market should aid it in assessing the impact on that public interest.

Other parties have also had to grapple with the appropriate definition of telecommunications markets, the DOJ being a focal point for much of that work. The DOJ follows a dual set of assumptions about the relevant geographic and product markets (see DOJ "Comments", 1984, pp. 13–15). From an economic point of view there is a single market which encompasses the national intercity (or interexchange) service markets. MTS/WATS, private line services, and public switched record services are all components of that market. This view is based on the DOJ assessment of both supply and demand substitutabilities. However, from a regulatory point of view DOJ recognizes that telecommunications markets are segmented into at least the interstate and intrastate service areas. The basic dichotomy is adopted for this study because, while in the process of on-going deregulation, telecommunications markets are not as yet completely free and open. This will remain true as long as the metes and bounds of where a carrier can operate are dictated by government regulation.

Arguments for Segmenting Telecommunications Markets. From the standpoint of the de jure and de facto market supply and demand it does not seem reasonable to combine the intra-LATA and inter-LATA markets. By the terms of the MFJ (1982) the BOCs cannot compete in the inter-LATA markets, and so cannot be considered as having relevant prospective supply. The same principle applies on the demand side. A buyer does not have the option to buy inter-LATA service from BOCs. As long as the constraints are legal rather than economic, it is not sensible to consider them part of the same product market. Thus, on the basis of present law a case can be made for treating the intra-LATA and inter-LATA markets as separate and distinct.

On the other hand, since the constraints on competition in the two markets are merely legal ones, things could change. Apparently most ICs are willing and able to service intra-LATA consumers, and most BOCs are willing and able to service the inter-LATA markets. Thus it can be argued that only artificial legal restraints keep the ICs and BOCs from head-to-head competition in virtually all markets. If the legal restraints were dropped, the economics seem to be in place for vigorous competition, both within and among LATAs.

Definition of Market Power

Unregulated Markets

Basic Definition. Traditional or standard definitions of market power have been developed in the economics literature and analyses by the FCC, DOJ and the Supreme Court. The Supreme Court has defined monopoly power as the power to control prices and to exclude competitors (American Tobacco, 1946; DuPont, 1956). The DOJ defines market power as "the ability of one or more firms profitably to maintain prices above competitive levels for a significant period of time" (DOJ "Guidelines", 1984, p. S-1). The FCC defines market power as the ability to raise prices by restricting output. A similar definition of market power has been discussed at length by Professors Landes and Posner (1981, p. 937). They state:

> The term "market power" refers to the ability of a firm (or a group of firms, acting jointly) to raise price above the competitive level without losing so many sales so rapidly that the price increase is unprofitable and must be rescinded.

In all of these discussions it is explicitly stated or implied that the exercise of market power would be profitable. The presumption is that an activity would not be undertaken if it were not expected to enhance the firm's long-run profitability. Therefore, unless a firm can profitably raise or maintain prices above the opportunity costs of its resources, it has no power in the marketplace. In essence, this definition implies that the exercise of market power would be expected to increase the present value of the firm.

Contestable Markets. Recent work by Baumol, Panzar and Willig (BP&W, 1982) on the contestability of markets emphasizes that it is entry, and the threat of entry, that constrains an otherwise dominant firm from charging a monopoly price or from exerting market power. Normally, economists think that to have vigorous competition it is sufficient to have: (1) free entry and exit with no sunk costs; (2) a large number of knowledgeable buyers and sellers; and (3) homogeneous (fungible) and easily substitutable products. However, a market is formally considered *perfectly contestable* if all potential sellers have the same technical and productive characteristics or the ability to acquire those characteristics, if buyers respond quickly to price changes, and if there are no sunk costs.[5] Under these conditions there will be no barriers to entry and prices will fall to average cost.

When the market is considered a natural monopoly, these same issues of entry and exit are described in what is called the sustainability literature (see Evans, 1983, Ch. 9). This literature examines whether a natural monopolist can ward off uninnovative entry; when it cannot, the market is considered *nonsustainable*. The tradeoff for regulators with nonsustainability is that they must prohibit entry to maintain the economies of a natural monopoly, yet they must forego the disciplining effects of competition. A natural monopolist who must accumulate capital in anticipation of growing demand is often operating in a nonsustainable market. For telecommunications it is especially noteworthy that any natural monopolist who faces a piggyback entrant—one who buys inputs from the monopolist and sells outputs that compete with one of the monopolist's offerings—has a serious nonsustainability problem.

Although BP&W's work is being challenged by writers such as William Shepherd, much of the criticism is about semantics rather than substance. The central message of much of this work is that the more important questions that need to be addressed concern the conditions of entry and exit in an industry, not the size or shares in a static market.

Market Power and Behavior of the Firm. From a legal point of view, possession of market power is not illegal per se. But the acquisition and/or maintenance of market power by illegal means does constitute an antitrust violation (Shooshan, 1984, p. 49). Thus, in addition to the existence of market power, some bad acts or conduct, the purpose of which is to obtain or maintain a dominant market position, must be found. This means that even if market shares and all the other market power indicia are present, no violation has occurred unless anticompetitive behavior is also evident. This anticompetitive behavior can take many forms, such as exclusionary pricing.

Although market power is defined in terms of the prices charged, its exercise may become manifest through inefficiently high levels of investment designed to limit entry. While these investments may be unprofitable in the short-run, they may enhance long-run profitability. Recent research has emphasized the ability of firms to use "strategic behavior" as the basis for the exercise of monopoly power (Dixit, 1982.

Market Power in Semi-Regulated or Asymmetrically-Regulated Markets

How should the presence of market power be determined in telecommunications markets? As a result of several forces, formal regulation of entry and pricing in the inter-LATA telecommunications market is being replaced by a system of free and open competition. Indeed, since the first OICs entered the national scene in 1978, competition has grown dramatically so that the market now has 400 vendors, with AT&T-C facing rivalry in every type of service offered including voice, data, and private line. Yet AT&T-C, alone among many OICs, remains subject to significant regulation, including the regulation of its tariffs and service offerings. Moreover, in Georgia, as well as nationally, roughly 60 percent of AT&T-C's interstate revenue will be paid to the local telcos in the form of regulated access charges, including billing. Of course, the efficacious management of these access costs is essentially out of AT&T-C's control since it no longer owns the local BOCs. While OICs also pay access charges, they have been deliberately set at significantly (up to 70 percent) lower level with the intentional design both to compensate for unequal access and to handicap the early track leader AT&T-C. The OICs will probably seek and maintain this imbalance even after there is complete equal access implementation (scheduled for 1986).

The semi-regulated or asymmetrically-regulated interexchange telecommunications market registers the year 1982, because of the MFJ, as a watershed. The framework is in place for vigorous competition in which efficiency dominates. But before that can become a reality, the regulatory agents must eventually remove all the bridles. For that to be feasible, no one carrier or group of carriers can possess significant market power.

We will next examine the factors that a trier of facts would examine to make that judgement in an informed fashion. For the moment, we merely wish to emphasize that as long as the interexchange market remains semi-regulated or subject to asymmetric regulation, the *dynamic* factors perforce must be given greater weight than the static factors (see Garfinkel, 1983).Almost any market moving from more to less regulation will contain firms whose current market shares are unreliable indicators of current or future market power.

Definition of Market Shares

As explained before, the Supreme Court and federal regulatory agencies have defined monopoly power to mean the ability to control prices and to exclude competitors. However, since the indicators of market power are difficult to establish directly, the courts often look to market share as the principal sign of monopoly power. Moreover, this has been a long-standing policy which continues to be practiced in the courts and in many regulatory tribunals. Therefore, the analysis of market shares is a starting point in nearly any study of market power.

Market versus Industry Distinction

The term market has already been defined as consisting of a set of suppliers, each of which is supplying products that have a significant degree of substitutability, to the same potential buyers. In other words, a market groups together *both* buyers and sellers who communicate and transact for a given set of goods or services. We have already shown that a market has at least three significant dimensions: product, geography, and time.

A market is not necessarily synonymous with the more commonly used term *industry* because firms in the same industry may not supply substitutable products and/or may sell their products to quite distinct groups of customers. Moreover, some firms that are perceived to be in a particular industry may produce many other products besides the one that led them to be so classified. For example, while Westinghouse Electric Corporation is normally considered as a supplier of electrical goods and equipment, it also produces gas-fired incinerators, plywood, and television programming. Put differently, an *industry* usually refers just to the suppliers and not the buyers in the market and often has many markets within it. As another example, the "steel industry" refers to the steel firms who supply in a variety of markets for various steel products in various geographical areas.

Market Shares

A *market share* refers to the percentage of a market supplied (or controlled) by a particular firm during a specified time period. This is usually calculated in terms of the firm's share of the total market's sales revenue or value added (in monetary terms) or of the output, volume, or capacity (in physical terms). For instance, if Firm A accounted for $40 million of the total sales of $100 million in the market during a year, its market share would be 40 percent.

Sellers' Concentration

Economists and lawyers in assessing market power are often concerned with the level of sellers' concentration in a market. Sellers' concentration refers to the distribution of market shares among the different member firms. This sellers' concentration can be calibrated on the basis of the distribution of revenues, assets, employees, or other relevant variables. The most common or popular measure of market concentration is the so-called *four firm sellers' concentration ratio* which shows the percentage of total sales (physical volume, value added, capacity, and so on) accounted for by the four largest firms. For instance, if the four leading firms sell $80 million out of total market sales of $100 million, the four firm sellers' concentration ratio is 80 percent.

Relations between Market Shares and Market Power

General

It is crucial for all interested parties to recognize that while market shares can be an important ingredient in estimating market power, they are not the same thing. This is true even in unregulated markets or industries. A large market share can be a result of serving the buyers better. If an unregulated firm can lawfully decrease its costs and prices and thereby obtains a larger market share, it should not be condemned as an antitrust violator. More generally, it should always be open to a firm to rebut an inference of market power based on market shares by showing that its market shares are the result of low (but not predatory) prices. It should also be emphasized that large market shares are the result of *past* economic conditions which may be forever gone and are in no way a measure of the state of future economic conditions and hence future market shares. Clearly, it is incorrect to infer the existence of market power simply from observing market shares.

Specific to Regulated Telecommunications Markets

The general statements in the preceding paragraph apply even more so to a regulated market. In fact, one of the theses underlying this study is that effective rate regulation makes market shares even less reliable surrogates for market power than they are in an unregulated environment. Let us examine this thesis and the related issues more carefully.

First, it should be almost self-evident that AT&T-C's large market share is the result of *past* economic conditions. AT&T-C's current market share is but an artifact of an earlier era when one enterprise served the entire market because the FCC and state commissions at first prohibited and later severely limited competition. If one accepts that the interexchange telecommunications market is contestable, is not a natural monopoly, or is without significant entry barriers, AT&T-C's current market share is likely, on balance, to dwindle over time.

Second, as explained by Landes and Posner (1981, pp. 975–76), regulation may enhance a firm's market share in situations where there is only the appearance and not the reality of market power. Thus, AT&T-C's current market share reflects in part its function as a "carrier of last resort" serving geographic areas and categories of customers that other interexchange carriers (OICs) have chosen to ignore. AT&T-C provides this service at rates that are marginally profitable at best, because of its nationwide average rate structure. In other words, regulated firms are often required to charge identical prices in different product or geographical markets, despite differences in the costs of providing the services. As a result, prices may exceed (marginal) costs in some markets and be less than (marginal) costs in others. In the below-cost markets, the regulated firm is likely to have a large, perhaps even a 100 percent market share.

The reason for the large market share clearly is not that it has market power, but that the market is so economically undesirable that the only firm that will serve it is one that is either forbidden by the regulators to leave the market, or enticed to remain in it by the opportunity to cross-subsidize and recoup its losses in protected and profitable markets. In these situations, a large, perhaps even a 100 percent market share is a symptom of a *lack* of market power.

In this situation the usual presumed causality between market shares and price is reversed. Instead of a large market share causing prices to be high, a low price causes a large market share. While the problem of reverse causality is not limited to the regulated industries, it has its greatest potential there. For instance, OICs in Georgia have gained a significant and growing share of the high-volume, high-profit, largely metropolitan business market, whereas AT&T-C has no significant competition in much of the low-volume, marginally-profitable rural and residential markets.

Brief Conclusions

In general, it is believed that, the higher the sellers' concentration ratio in a market the greater the potential market power (that is, the incentive and opportunity for conspiratorial, cartel, or collusive behavior is generally greater). This conclusion is subject to several serious shortcomings, some of which are considered in this report.[6]

Market power is a function not only of market share but also of the potential supply from either existing firms or firms that could enter the market. Sellers' concentration merely reflects the number and strength of *actual* market rivals of a firm, whereas the conditions of entry inform us to the number and strength of *potential* rivals. Thus, a one-firm market sellers' concentration ratio could be 100 percent and yet the market power of that one firm effectively could be zero if the potential supply elasticity were great enough. In other words, a supernormal or monopoly price in this situation would cause the market to be deluged by expansion of existing (marginal) firms and by new entrants.

Factors to Consider in Assessing Market Power

In making an assessment of the market power of AT&T-Communications (hereinafter referred to as AT&T-C) it would be necessary to evaluate the probable demand responses of consumers and the probable supply responses of other firms. More specifically, the profitability of a price increase by AT&T-C will depend, among other things, on the responses of: 1) consumers who may reduce their usage, switch to other services offered by AT&T-C, or develop their own network; 2) consumers who may switch to comparable services offered by other common carriers; and 3) other common carriers who expand their presence in the market or who enter the market for the first time.[8]

Determining the probable responses may be helped by information and judgment about both the historical responses of consumers and potential competitors. It is also often instructive to render an assessment of the regulatory environment and the changing state of technology. However, the emphasis should not be exclusively on historical relations, but on the probable future responses of consumers and alternative carriers.

It should be noted that the following list of the factors influencing market power is at times redundant. For example, the influence of product differentiation on market power is investigated both as an element affecting entry conditions and as a separate influence. We believe these overlaps are instructive because various sources have considered these factors at different levels of aggregation. We prefer in this study to err on the side of excessive analysis in total, rather than risk a deficiency in any particular area of investigation. We hope that the repetition and the dovetailing that is involved with this format is neither too tedious nor too abstruse for the reader.

As we have already suggested, market power is a mosaic and each of the factors we have set out is a piece in that mosaic. Because this study focusses on trying to develop a methodology for monitoring market shares, we will examine them first. However, we attach no significance to the ordering or to the length of our discussion of each of these factors. A systematic and valid final assessment of market power would require an examination of at least these factors. Then, on the basis of an as yet not universally agreed upon weighting system, a decision could be rendered. Economists, to a certain extent, and lawyers, judges, and regulatory agencies to a substantially greater extent, have historically placed greater emphasis on market shares than any other factor—perhaps, some critics would say, because they are quantifiable. This emphasis is changing rapidly. The more enlightened assessors of market power—whether communication leaders, economists, judges, lawyers, politicians, or regulators—are much more involved in the type of schemeta set out next, although we are unaware of any who use all of the factors we suggest. Rather, they truncate the list, either because of some of a methodological preference, or because the categories overlap, and typically focus on only five or fewer of these factors discussed below. We do not mean to imply that we have all the answers or that the received economic doctrine as presented here is the best approach possible; but in the absence of any superior approach we must use what we have, and give it up only when improvements occur. In any case, we can think of only two alternatives to using economic theory if we want to know the impact, say, of a new type of regulatory environment. They are divine revelation, about which we have nothing to say, and historical experience or experiment, of which there may not be any.

Level of Sellers' Concentration or Level of Market Shares

In general, the higher the firm's market share (all other things being the same) the more likely it is to have market power. But as we have already indicated, this is subject to a host of provisos and qualifications.

We have already discussed the limitations on using the level of market shares or the level of concentration as the sole measure of market power. In general, we will not repeat them here. Despite the limitations of market shares as a *monolithic* measure of market power, almost everyone agrees that it would be worthwhile to examine their level and trend. For instance, Lexitel Corporation ("Comments", 1984, p. 21) intoned that "the Commission should establish several measures of market share and monitor their relative change over time."

More Limitations

One point that we have not emphasized is that, according to some scholars, knowledge of sellers' concentration by itself does not enable an assessor to make reliable predictions about market power, *except when concentration is very low*. The likelihood of market power under medium or high sellers' concentration is affected by a number of misunderstood features of products, markets, and modes of contractual interaction. As a result, one study (Joskow and Schmalensee, 1983, p. 192) concludes thus: "As deregulated markets do not seem likely to involve levels of concentration that are so low as to compel competitive behavior, an inquiry into other factors and influences on market conduct is required."

There is one final point that also merits our consideration. It may be that a high level of concentration that follows in the wake of rising market shares may reflect heightened competition. There are certain well-known efficiency distortions that flow from regulation (such as the Averch-Johnson effect). But there are other distortions that may have resulted in the previously regulated telecommunications industry, such as the mandate that AT&T obtain the FCC's approval of its constructions plans, limitations on AT&T's ownership and use of satellite technology, and the practice of nationwide rate averaging. These regulatory requirements may have resulted in an inefficient investment program, thereby putting AT&T at a relative disadvantage as compared to its rivals. The move toward dismantling of regulatory controls may allow the less bridled growth of a more efficient AT&T-C. As Besen and Woodbury (1983, pp. 61–62) put it:

> Even if this [heightened efficiency] were to result in a continued high market share, and possibility market power, for AT&T, Williamson argued some time ago that even modest cost reductions may be large enough to outweigh the loss of consumer welfare that might stem from that market power.

Traditional regulatory enforcement often considers the analysis of market shares as the starting but never the final point in evaluating the competitiveness of a market. But as Haring (1984, p. 18) indicates: "Modern analysis strongly suggests the one *cannot* unambiguously equate declining market shares with improved industry performance, particularly when market shares are determined in whole or part by regulation (or strategic responses to regulation) rather than by competition."

Evidence in Georgia

In the next section we indicate what we do know about levels and trends of market shares in Georgia given the paucity of the available data and hence we do not repeat it here.

Trend of Concentration or Trend of Market Share

General Considerations

In general, the greater the ability of the leading firm(s) to maintain market shares over time, (everything else remaining the same) the more likely it is that market power is operating. The assessment of market power in a previously regulated market is different from such an assessment in an unregulated market; the rate of decline in market power is of expanded importance vis-a-vis the presence or absence of market power, as the latter may have existed previously by deliberate regulatory design.

Unless a previously regulated firm has some cost advantages over actual and potential entrants, it will not be able to raise prices above costs without attracting a host of entrants whose increased production will depress price back to cost. Thus, even a firm with 100 percent market share is subject to the competitive discipline of potential entry, and therefore the chilling winds of competition.

AT&T-C's declining national market shares in equipment and long-distance service markets is a possible indication that even if AT&T-C had been a monopolist at one time, it may have lost the power to raise prices and to exclude competitors and is thus no longer a monopolist. In the Alcoa case (1945), for example, declining market shares of the monopolizing firms were a major factor in determining that divestiture was unnecessary.

While rising market shares are a theoretical possibility for AT&T-C if the removal of some of the regulatory schackles lead to a sufficiently greater degree of efficiency, we do not give this issue much practical importance. The reason for this assessment is that AT&T-C still remains with a significant degree of oversight by regulatory agencies. In addition, we believe that there is a substantial degree of competition from the OICs, resellers and bypassers, that largely negates any likely efficiency gains by AT&T-C resulting from its quasi-deregulation.

What has happened to statewide market shares in Georgia since the MFJ? Although there is no definitive study on this, there is modest evidence that OIC's have had considerable success in penetrating high volume markets and eroding AT&T-C's market share with heavy users.

Number and Type of Sellers

In general, the more actual and potential sellers there are in a market (including their internal or captive consumption, if any), everything else the same, the less likely market power is present. There has been significant, substantial, and continuing entry into the toll market in recent years, both nationally and in Georgia. There are a legion of actual and potential sellers of intrastate, inter-LATA telecommunications. After naming these different types, we shall briefly discuss each of them. The types include:

(1) facility-based competing long-distance telephone companies; (2) satellite carriers vis-a-vis terrestrial carriers (for the former costs are basically distance insensitive); (3) resellers; (4) private telecommunication networks—so called "bypass" occurs when a service is originated or terminated without using the local exchange carriers; (5) others—railroads, existing local telcos;

larger users; high technology equipment manufacturer; and AT&T-Information Systems.

Let us examine each of these in more detail for the state of Georgia. Our discussion is at best probative and not dispositive, for the focus of this study is on methodology and not on statistical results.

Facility-Based Inter-LATA Common Carriers in Georgia

At the end of 1984 there were three facility-based common carriers authorized to provide inter-LATA long distance service in Georgia. The largest was of course AT&T-C, with MCI Telecommunications and GTE Sprint in the second and third positions. A fourth, Microtel, had certification pending. However, service has been provided without authorization in some places, and there are potential entrants who presently offer interstate long distance service, including United States Transmission Systems (USTS), a subsidiary of International Telephone and Telegraph Company (ITT); Western Union; and U.S. Telephone, recently acquired by United Telecommunications. Each of these has a long history of providing telecommunications services and is a potential entrant into the Georgia inter-LATA market.

Facility-Based Intra-LATA Long Distance Carriers in Georgia

The almost 40 existing local exchange companies have experience in offering short-haul toll services and could become viable competitors. Bell South and Southern Bell are worthy competitors in the intra-LATA long-distance market and they can be expected to seek persistently entry into the inter-LATA market. The roughly 10 companies that are authorized by the Georgia Public Service Commission to offer originating service in Georgia need also to be considered. These companies are also authorized to carry intra-LATA toll traffic.

Resellers

Resellers got started nationally in 1980, essentially reselling volume discount MCI telephone service. They took advantage of the Execunet volume discounts to resell MCI services in a similar fashion as AT&T-C's private lines and WATS are resold today. The big breakthrough, however, came in 1981 when the FCC ruled that all intercity voice services could be resold including private lines, WATS and MTS. From a trickle of revenue in 1981, the business exploded in 1983 reaching about $800 million annually. Most of this revenue was created by the top five resellers, which nationally in 1982 were U.S. Telephone (three times the next largest reseller), Combined Network,

Satelco, Lexitel, and Teltec. TeleStrategies projects that by 1988 this figure should grow to $5 billion.

Telecommunications resellers may be classified as primary or secondary resellers. The former are companies whose main line of business is that of reselling the services of facility-based carriers. Secondary resellers, on the other hand, are mainly engaged in some other line of business, but have spare capacity that can be profitably resold. Secondary resellers may have constructed some facilities of their own or be buying from facility-based carriers at bulk rates. Examples of secondary resellers include financial institutions and insurance companies, hotel chains, and other large businesses with diversified locations. Since resellers are not required to be certified, it is difficult to know with certainty how many are presently operating or how much of an impact they are having on the net revenues of the facility-based common carriers. However, the primary resellers known to be operating in Georgia include Allnet (Call America), Central Telecom (Touch and Save), Isacomm (Network I, Inc.), Pacific Atlantic (Satelco), Starnet (Trans-call America) and TDX Systems, Inc. (TMC). Typically, they purchase WATS lines from AT&T-C and/or equivalent services from other facility-based carrier, provide switching and billing services, and resell at rates below the MTS schedule. They sell to both business and residence customers. Whether resellers are primary or secondary, they represent potential competition for the facility-based carriers.

It has been suggested that in a deregulated environment, ICs could drive resellers from the market by raising the price, for example, of volume discount services. Although this is possible, it does not seem to be economically logical. AT&T-C, the largest supplier of service for resale, would probably determine that such a pricing strategy is not profitable. The only sensible reason AT&T-C would want to drive resellers from the market would be to precipitate a switch of those customers to AT&T-C. But this would not be likely to happen. Reseller subscribers use this service because of its lower prices and/or because they do not need or care for the offerings provided by a higher-priced, full-service carrier such as AT&T-C. Therefore, if resellers are driven from the market, most of their subscribers would *not* be driven to AT&T-C, but to the OICs, who offer the lower priced, no frills service those customers desire. Since the OICs generally tend to be more formidable rivals than the resellers, AT&T-C would not benefit by driving the previous resellers' subscribers to the OICs. In fact, diverting resellers from the market could cost AT&T-C a considerable sum. Both state and federal price discrimination laws preclude firms such as AT&T-C from pricing services such as WATS differently to different customers. Thus, in order to drive resellers from the market, AT&T-C would have to raise its WATS prices to all customers. This pricing move could decrease AT&T-Cs revenues by shifting many traditional WATS users to OICs or private networks.

Private Facility-Based Carriers

Private telecommunications networks are also part of the competitive marketplace in which AT&T-C operates. To the extent that customers provide their own transmission systems and networks, they divert traffic and revenues from the common carriers. Users in both the business and government sector with large volumes of voice and data traffic are easily able to develop strong cost justification for providing their own systems. The low marginal cost of these private microwave systems encourages their operators to resell excess capacity to other users, thus further competing with the traditional carriers. For example, the capacity of private microwave systems has been growing significantly. In 1974, there were about 54 million private microwave circuit miles. By 1978, this grew to 68 million miles, and then to 107 million miles in 1983.

Other Potential Competitors

Other potential competitors who may not ordinarily be considered to be in the interexchange market include companies that are large users of information transmission networks and who already compete in the information services market. These include major financial firms, remote computing services companies, television networks, and value added carriers as well as the private microwave/private system operators. One such group of potential rivals whose supplyside substitutability is probably quite high, includes railroads and other firms that already have rights-of-way necessary for fiber optic networks and similar facilities. In the past such companies have entered the market and appear poised to do so in the future. For example, Southern Pacific founded SPCC/Sprint, which is now a major provider of interexchange switched service through GTE-Sprint. More recently, AMTRAK, CSX, the Florida East Coast Railway Company, and the Missouri-Kansas-Texas railroads have announced joint ventures for the sale or lease of their rights-of-way to codevelopers to establish fiber optic networks.

Number and Type of Buyers

In general, the fewer and the larger the buyers the more likely countervailing power will prevail and the less likely market power can operate. Large buyers include both the traditional high volume customers and resellers. Both have credible threats they might use to countervail any potential market abuses by sellers. They also have the opportunity to switch to other ICs as well as the possibility of establishing their own facilities.

While we will not have any specific data on high volume users (i.e., potential bypassers) in Georgia, as information could be obtained with a modest effort. We have already discussed what information we have on resellers in Georgia.

Barriers to Entry and Exit (Conditions of Entry and Exit)

It is at least as important to assess the conditions of entry (Bain absolute or Stigler relative variety) and exit in determining market power as in determining the level (and trend) of market shares.[9] For instance, a high market share without legal, technical, financial, or economic barriers to entry does not necessarily connote market power. Thus, the importance of considering the conditions of entry in assessing market power can not be overemphasized, for as Hal R. Varian (1984, p. 91) states: "If the monopolist is to remain a monopolist, there must be some sort of barrier to entry." In general, the more imposing are the conditions of entry, everything else the same, the more likely it is that market power can be exercised. Conversely, while we know that Adam Smith's invisible hand forces businesses to serve the greater good, "this hand moves most swiftly when businesses are many, and the least-cost scale of operation is small relative to the size of the market, and there is no barriers to new businesses" (Evans, 1983, p. 63). While many of our comments relate to the barriers to entry at the national level, it is also important to determine either qualitative or quantitative state specific conditions of entry.

Barriers to Entry in General

The term *barriers to entry* refers to obstacles preventing new firms from engaging in the production of a particular category of output. It is important to keep in mind that there is a divergence of opinion on what constitutes a barrier. The older approach is that of Joe Bain.[10] He defines a barrier to entry as the ability of established firms to elevate prices above minimal average costs without inducing new firms to enter; or, more simply, barriers to entry are any costs that a new entrant must bear that an incumbent does not have to bear. Examples of the latter include the availability of cheap capital to larger firms (imperfect capital markets), scale economies, learning curve effects, fixed costs, certificate-of-need requirements imposed by regulators upon entrants, and lack of access to bottleneck facilities. Followers of the newer so-called Chicago School, championed by George J. Stigler, only consider a barrier to entry to exist when the relative costs between incumbent (existing) firms and potential entrants differs. Put differently, Bain argues that barriers to entry are any advantage that existing firms have over potential entrants, whereas Stigler contends that barriers to entry are only those costs of producing, for any given level of output, that must be borne by new entrants but not by existing firms. Thus, for instance, Bain considers economies of scale a barrier to entry, whereas Stigler does not because such economies are available to all. In the Stigler approach there is a distinction between reversible and irreversible investments. Fixed costs that require merely reversible investments do not pose barriers to entry. Fixed costs that require irreversible investments may pose barriers to entry. In the Bain approach, extensive fixed costs for entry create barriers to entry which impede competition. Since the Bain approach still predominates in the regulatory tribunals and the courts, if not in the industrial organization literature, our presentation is cast as a variant of the traditional mold. Thus, barriers to entry will be presented under three classifications: absolute cost advantages, economies of scale and scope, and product differentiation.

Absolute Cost Advantages. An absolute cost advantage may be due to patent protection, competent management, possession of strategic raw materials, vertical integration, and the availability or the low cost of capital.

Economies of Scale and Scope. The meaning of economies of scale for a single-product firm is well established. Formally, it describes the technical relation between physical input and physical output. However, assuming input prices are constant, there are economies of scale if output increases proportionately more than costs as greater quantities of the single product are produced. Moreover, it is a "static concept" in the sense that the state of technology is assumed to be fixed.

In other words, economics of scale refer to the basic cost conditions in the market, or, more specifically, to the cost reductions that are brought about by an expansion in the scale of operations of the sole product or service offering. Where there are economies of scale, the unit costs of solo production decline as the level of output increases. Graphically, economies of scale are represented by average cost curves that decrease as production levels rise.

In a multiproduct context, however, there is no single, meaningful measure of output or of average costs. Aggregating the products results in the loss of information about each product's specific effects on costs and how each product is affected by the mix and level of all the other products being processed. To capture the effects of different output level combinations in multiproduct applications, two measures of scale economies are used: ray (or overall) economies of scale, and product-specific economies of scale. Ray economies of scale measure the response of costs to the same proportionate increase in all the products: that is, the output composition is held constant while the scale is varied.

Product-specific economies of scale measure the behavior of costs as one output level is changed while the other output levels are held constant. Although scale economies show the effect of different output levels on costs, they cannot show how different product line combinations affect costs. A measure which does indicate the effect of different product line combinations on cost is called economies of scope. In the two product cases, economies of scope exist if the costs of producing the two products jointly is less than the total cost of producing these two products separately.

Why should a single firm be able to produce a given level of output of each product line more cheaply than a combination of separate firms, each producing a single product at a given output level? Bailey and Friedlander (1982, pp. 1024–1048) suggest three sources of scope economies: shareable or jointly utilized inputs, networking, and shareable, intangible assets. Shareable inputs arise in instances of joint production (beef and hides, wool and mutton, etc.) and production with a fixed factor with excess capacity (railroad trackage for both freight and passenger services). Network economies are thought to exist in the aviation, trucking and telecommunications (both transmission and switching networks) industries. Finally, scope economies may arise when the costs of sharing intangible assets (such as research and development activities) and business expertise between divisions in one multiproduct firm are lower than the transaction costs that separate firms would incur by trading this information.

The concepts of product-specific scale economies and economies of scope together produce a measure of multiproduct scale economies. Clearly, changes in the composition of output can have two effects. One possibility is product-specific decreasing returns to scale, but economies of scope as the costs are spread over more products. Thus, a cost function of a multiproduct firm is sensitive both to the composition of output and the scale of output.

As a firm alters its level of output and its product mix, costs may change in a variety of fashions, perhaps exhibiting both economies of scale and scope at one level of output, but exhibiting only scope economies concomitant with product-specific scale diseconomies at some other level. While there are relatively few empirical studies of multiproduct costs functions, the rather limited evidence strongly indicates that seriously biased estimates may result if the multiproduct nature of output is not explicitly considered. Whether AT&T-C is a single product or a multiproduct firm *in certain markets* may be contested, but at this stage the preponderance of evidence seems to suggest it is a multiproduct firm.

Product Differentiation. Finally, product differentiation refers to the perception among consumers that two products differ from each other enough to warrant a price difference between them. One product may be regarded as superior to another because the firm is more favorably located, the financial terms and conditions more attractive, the service and advertising better, or the product quality higher. Let us examine each of the three (Bain) classifications of entry barriers in more detail for the telecommunications industry.

Barriers to Entry in Telecommunications

Absolute Cost Advantage. Absolute cost advantages which can cause entry barriers include a variety of factors like patent protection, competent management, possession of strategic raw materials, vertical integration, availability and cost of capital, and so on. After divestiture, two alleged causes of market power based on absolute cost considerations were removed or at least reduced: (1) The "bottleneck" theory that AT&T's ownership of local exchange facilities could result in closed interconnection or discriminatory interconnection against competing interexchange carriers was eliminated; (2) The size theory (that AT&T's sheer size caused domination) was removed, as AT&T-C is now less than one-fourth of its former size and faces rivals of comparable size and financial strength such as IBM, ITT and RCA.

In addition, with deregulation another cause of market power was reduced: the advantages of AT&T-C trying to indulge in legal predation or investment in barrier-to-entry capital of a political firm were reduced. A rational regulated firm may find it good strategy to invest in both political activity and productive activity. The former includes all the techniques of obstruction of competitors available through the political process. Because the Noerr-Pennington (1961, 1965) legal doctrine, derived from Supreme Court rulings, shields businesses from liability for participating in governmental procedures (unless for the sham exception), it was claimed that AT&T-C previously had tried to use the regulatory process to impose sunk costs on the entrants, thereby discouraging entrants and encouraging exit.[11] There are some other bits of evidence which, while not apodictic proof, suggest that the absolute cost of entry is not insuperable in the intrastate inter-LATA markets. First of all, OICs such as GTE Sprint or MCI have indicated in South Carolina and elsewhere that there are low capital requirements necessary for entry into the toll market. In addition, besides the potential entry of OICs, many large companies have set up their own telecommunication systems which can be used as the basis for expansion. Moreover, cellular radio and cable TV systems can be rather quickly, easily, and inexpensively extended to enter the traditional toll telecommunications market. Further, it is particularly important to realize that the capital costs of reseller competition compared to transmission carriers is especially low running, perhaps in the range of $25,000–$500,000.

Economies of Scale and Scope. These are the cost reductions due to greater output or a larger product line. Competition in an open, unregulated market cannot endure if there are significant economies of scale. Whereas before divestiture, some econometric and engineering research on Bell System costs had found scale economies across a broad range of long-distance operations—particularly over the major trunk routes and in "networking"—the cost conditions are uncertain today. The DOJ ("Comments", 1984, p. 10, 39), for instance, believes that long-distance telecommunications are no longer a natural monopoly.[12] We are not aware of any hard post-divestiture evidence supporting the claim that AT&T-C is more efficient than specialized suppliers because it has economies of scope arising from its use of shared inputs to produce several outputs, nor that it has economies of scale arising from aspects of network planning.

According to Haring (1984, pp. 8–12; also see Kelley, 1982) there are methods for determining whether the telecommunicators cost functions exhibit subadditivity: (1) econometric estimation of cost functions; (2) analysis of market shares; and (3) test of open market competition. The problem with econometric estimation is that it is dauntingly difficult and often impossible to ascertain within the present state of the art if global rather than local subadditivity is present. Moreover, econometric authentication has been bothered by conflicts in the actual estimates that have been attempted. Regarding (2), in regulated markets the significance of market shares as a surrogate of market power or competitive effectiveness is attenuated. Because of the perverse outcomes that can result from strategic behavior, such as signaling problems in response to judgemental criteria, levels and trends of market shares are not failsafe measures of competitive effectiveness. As to (3), ultimately the only way of determining if telecommunications now or in the future is in fact a natural monopoly is to allow free competition, which is in the end is an evolutionary process of discovery and selection.

Product Differentiation. A barrier to entry may be due to such factors as superior location, favorable credit terms, persuasive salespeople, imaginative advertising, variations in product quality. We have some limited information on this element both nationally and in Georgia. We know that nationally, rapid progress toward equal access for all carriers has reduced this impediment to competition. Although the precise interpretation of equal access in the MFJ has been debated, we believe that the general thrust is clear: each BOC is required to provide to all interexchange carriers and information service providers exchange access and information access, and to exchange services for such access on an unbundled, tariffed basis (that is, equal in type, quality, and price to that provided to AT&T and its affiliates).[13] The equal access obligation took effect on September 1, 1984, and it will be phased in over a two-year period. By September 1, 1985, the BOCs had to provide equal access to one-third of their respective access lines, and by September 1, 1986, to the remaining access lines.

A number of BOCs proposed to connect OICs to the homes or businesses of local telephone subscribers by using facilities built or upgraded specially for that purpose; AT&T-C, on the other hand, will in numerous LATAs gain access to subscriber lines through different facilities, many of which are already extant.[14] For that reason, the access provided OICs will not be identical to that given AT&T-C in all places; rather, it will be technically different in certain areas and at certain times. The BOCs assert that any such technical deviations will be so slight as to be imperceptible to all customers, whether of voice or of data. It was asserted that even the interexchange carriers would not be able to detect any qualitative difference. Accordingly, they urged the Court to accept a definition of "equal access" as access whose "overall quality in a particular area is equal within a reasonable range which is applicable to all carriers," and to reject a more stringent definition which would demand access that yields identical technical quality (that is, identical values for loss, noise, and echo, and identical possibility of blocking).[15] The Court accepted the BOCs' definition and indicated that it would not insist on absolute technical quality. It reasoned that to do otherwise would necessitate substantial dismantling and reconstruction of local telephone networks, without any real benefits either to the consuming public or to the OICs. This ruling, however, was based upon the BOCs' representations that both voice and data customers will perceive no qualitative differences between AT&T-C transmissions and those of its competitors—at least with respect to those portions of the transmissions carried by BOCs.[16] The BOCs are well aware, however, that the DOJ will be monitoring the quality of the access they will provide and that the price of falling short of their assurance of equal quality will be enforcement proceedings brought by the DOJ.

With equal access, telephone subscribers are able to select or "prescribe" to a preferred OIC. Where available, calls dialed with "1" followed by the area code and telephone number are routed automatically to the preferred carrier. A five digit code can be used to access ICs other than the preferred carrier. Before conversion to equal access, OICs customers must dial 22 digits and have touch-tone service or use touch-tone pads. The parity will allow a unique marketing opportunity for OICs since all users will be asked to specify some inter-LATA carrier it requires them actively to reconsider their carrier. On the state level, surveys of perceived quality differentials suggest that AT&T-C's product service offerings are not as great as the present discounts on access charges. However, it is true that AT&T-C is the only interstate carrier with POPs in each of the 161 LATAs in the United

States, although the OICs can terminate traffic in all LATAs through resellers.[17]

Two Views

Some argue that barriers to entry that favor AT&T-C over OICs include first-mover advantages, large capital requirements, scarce know-how and inputs, difficulty of upsetting established customer connections, absence of a known performance record, and the inability to exhaust the bountiful economies of scale that are possible.[18] It is further claimed that AT&T-C can discourage entry or hinder intermarket mobility because it is the sole interexchange carrier with known premium access. It has a more extensive network; it retained the efficient Western Electric and imaginative Bell Labs; and it obtained an embedded base of CPE and a 50 percent share of the CPE market. We were unable to validate or refute, on the basis of our casual empirical work, many of these allegations and "facts." A more careful investigation is needed before actual empirical estimates of market shares and market power in Georgia is to be circulated. However, the substantial increase in the number and size of firms providing interexchange telecommunications service in recent years, both in Georgia and nationally, is evidence that barriers to entry are not insurmountable.

Product Differentiation

Some feel that product differentiation is so significant as a potential factor in market power as to be out by itself, in addition to being discussed as a factor influencing barriers to entry. In general, the greater the degree of product differentiation, everything else the same, the easier it is for a firm to acquire market power. Product differentiation exists if the products of different firms in the same market are not perfect substitutes for each other from the point of view of buyers. Product differentiation can also occur because one firm has a more extensive product line. For instance, telephone companies in addition to their *ordinary services* (residence, public telephone, mobile, and basic individual-line telephone services) offer about twenty-five categories of *special services or specials* (foreign exchange, WATS, PBX, Centrex, private-line, and private network services) which require special treatment with respect to transmission, switching, billing, or customer use, and are mainly used by businesses. The overall demand for these special services is growing about twice as fast as the demand for ordinary services. When there is significant product differentiation, the absence of a well-defined product market (how do you aggregate apples and elephants?) may cripple the reliability of any market share calculations, and one may need to look elsewhere to help with determining market power.

The Equal Access Issue

It has been argued that the OICs have a product differentiation limitation related to the availability of equal access. While parity is projected to be attained by 1986, equal access to local switching facilities is simply not universally available at present. Therefore, the OICs must settle currently for an interconnection arguably inferior to that of AT&T-C. For instance, in most cases consumers of the OICs' services must dial more digits than users of AT&T-C's services and they must use touch-tone phones or pads. If the services of the OICs are perceived—whether correctly or not—as less convenient than AT&T-C's or of lower quality, they are less than perfect technical substitutes for AT&T-C's services and thereby represent less of a disciplining factor on AT&T-C's ability to raise the price of its services above competitive levels.

Of course with the advent of parity, the inherent disparity in convenience and quality between the services of the OICs and AT&T-C will diminish. However, the OICs may decline the enhanced services due to cost considerations. The MFJ required that the BOCs begin offering equal access in September of 1984 and that they make it available on demand by September of 1986. While some have thought that the nonavailability of equal access in a significant number of exchanges will probably impose restraints on the competitiveness of the OICs for the next few years on a product quality basis, others suggest that the perceived product quality differentials can be and are being overcome by price discounts. Clearly, in Georgia, MCI and Sprint have achieved relatively high levels of brand recognition as residential customers have heard or read of MCI and Sprint as long-distance carriers.

Moreover, the OICs are presently spending considerable amounts of money on advertising. For instance, in fiscal 1984, MCI spent $60 million on advertising nationally (compared with $29 and $21 million respectively in fiscal 1983 and 1982). In addition, OICs such as MCI have already copied AT&T-C in terms of organizational structure, information service, directory assistance, and travel cards.

According to Salomon Brothers (1984) MCI is making great strides in narrowing its product differentiation disadvantage based on limited accessibility. The growth of this second-largest long-distance carrier has been striking, with fiscal-year expenditures of $272, $623, and $889 million respectively in 1982, 1983, and 1984.

The gross investment in MCI's communication system now exceeds $2.5 billion with annual spending on construction in 1984 at about $1 billion with $1.1–$1.2 billion expected in each of the following years. Its construction priority is to add capacity for both the voice and data services traffic expected by equal access. MCI is adding about 2,500–3,000 local access lines every month and will maintain or accelerate this pace over the next few years. MCI expects to establish a POP in every LATA by March, 1986. MCI is also seeking approval to carry intra-LATA calls, with no rate regulation and mirroring of interstate access charges. About half of the states that have considered allowing intra-LATA competition are permitting companies other than LECs to carry toll traffic. It also expects to gain inter-LATA certification in each state by the time equal access becomes available.

Evidence in Georgia

Again, while we have limited information on this source of market power for the state of Georgia, our general comments—without other evidence forthcoming—are likely to apply to Georgia.

Technology and Dynamics

Generally speaking, the more technological activity, invention, and innovation that goes on in an industry, the less likely it is that any one firm or group of firms can acquire and utilize market power. Moreover, in a market with dynamic technological innovation, such as in the toll telecommunications market, existing market shares may not be an accurate surrogate for or index of market power, especially when this share was achieved when the market was closed to entry by regulation and that situation no longer exists. A concentrated market share can be eroded quickly when entry is relatively easy and entrants can focus on only the most profitable (typically the largest) buyers in the market. If one of the firms such as AT&T-C is subject to retail toll schedule regulation under a state-wide averaging mechanism, it will lose its winners and win its losers.

Technology in Telecommunications

Three Broad Trends. The recent rapid technological changes that have occurred in telecommunications have greatly contributed to the vigor of competition. (See Kaserman, 1984, pp. 140–144). There have been three broad trends that have substantially affected the potential for competition in this market.

First, profound developments in transmission technologies have greatly decreased the scale of economies in the provision of the interexchange telephone service. These technological developments, in conjunction with the rapid demand growth, may have eliminated the natural monopoly conditions that once characterized the industry (see Evans, 1983, Ch. 6, pp. 127–156, and Evans and Heckman, 1984). This means that a greater number of firms of minimum efficient scale can now coexist for any given level of market demand.

Second, technological developments have significantly decreased the costs, especially the sunk costs, of new firms entering into the toll telephone market. The opportunity of leasing satellite capacity substantially decreases the magnitude of the initial investment required to enter the market. In addition, the necessary ground facilities required to make this capacity operational are largely capable of being moved to alternative locations at a modest cost (see Katz and Willig, 1983). In short, the irretrievable sunk costs of entry have been greatly reduced through technological advances, making the market more contestable. Thus, potential competition can now operate even more effectively to restrain and discipline incumbent firms' economic behavior and performance.

Third, technological developments have widened the spectrum of consumer demands which in turn has increased the number of firms that can operate successfully in this market.

Technology and Competition

The authors of this chapter believe that it is virtually impossible to exaggerate the importance of technology to competition. There is perhaps no greater danger to market power than the "creative destruction" of technological innovation. In the words of Nourse and Drury (1938, p. 221).

> The man who today tries to fence in an industrial highway and exact an exorbitant toll from those who would travel this road to consumer satisfaction is in danger of defeating himself. Under modern conditions of technology, applied science is likely to find other means of progress. The chemist will build a detour around him, the physicist will drive a tunnel under him, or a biological overpass will be devised.

Thus, we must acknowledge the existence and interpret the likely consequences of singular technological developments. We know that dramatic technological progress has resulted in the availability of multiple technologies such as fiber optics, satellite communications and microwave transmission for AT&T-C's competitors to utilize. In fact, most technological change in the post World War II period has occurred in the long distance and traffic sensitive area through microwave technology, satellites, and greater multiphasing of channels. This cost-reducing technological progress has resulted in a greatly accelerated rate of entry and expansion. Toll service utilizing satellite transmission (provided by such companies as IBM, Mitel, and NCR) is projected to increase by more than 200 percent between 1982 and 1987, with AT&T-C expected to be operating less than 11 percent of the satellite capacity by 1987. Land-based transmission capacity of AT&T-C's competitors has also increased significantly by employing fiber optic technology. In the year 1984, AT&T-C's rivals will build over 11 thousand miles of fiber optic cable. This represents a staggering 95 percent of total route mile fiber construction. The microwave transmission capacity (as measured in circuit miles) of private bypassers has grown at exponential rates, doubling between 1974 and 1983. In 1982 OIC and private microwave capacity accounted for 72 percent of the total microwave route capacity.

Effects of Past Regulation on Technological Change

Despite the remarkable achievements which have in fact come to pass, there is evidence that regulation has tended to attenuate the returns to technological innovation in telecommunications, as with other public utilities, and thereby harm the dynamic efficiency of the industry. Thus, while the invisible hand rewards innovative entrepreneurs who discover and satisfy latent demand, in regulated markets the invisible hand may be jaundiced or all knuckles. For instance, the optimal timing of technology was stifled when Bell Labs was delayed ten years by regulation after developing cellular mobile phone service and AT&T was stalled in trying to bring "data under voice" digit communications capacity to the market.

Evidence in Georgia

While we have uncovered no comprehensive statistics on technological innovation accomplishments and potential for the state of Georgia in our cursory data investigation, we have no reason to believe that our general comments based on the national level do not apply mutatis mutandis to the state of Georgia.

Contestable Market Theory

In general, the closer a market is to being (perfectly) contestable, all other things the same, the more likely it is that firms will have no market power. The theory of contestable markets emphasizes that the rigid conditions of pure competition, particularly that there be a "large number" of firms, are unnecessary for society to reap the benefits of competition. The conductibility of a market is assured as long as entry is free and exit is costless, regardless how many firms are in the industry or how large the market share of any one participant. In short, a market is contestable if the threat of entry attenuates the ability of existing firms to charge prices greater than their costs. It is considered perfectly contestable if entrants can enter and exit costlessly when there are supernormal or subnormal profits. The smaller the sunk costs, and the more incumbents can respond to competitive price cuts, the more contestable the market. The concept of perfect contestability is even more general than that of perfect competition; the latter requires small and numerous sellers and a homogeneous product, whereas a contestable market does not. The freedom of entry and exit are sufficient conditions for contestability and economic efficiency. In other words, while a perfectly competitive market is necessarily a contestable market, a contestable market is not necessarily perfectly competitive.

More simply, effective competition requires that (1) consumers must be willing and able to change suppliers if there is a change in price, and (2) competing carriers must be willing and able to increase output to meet the increased demand if another carrier raises its prices above costs. Alternatively effective competition is said to exist when users have the ability to obtain comparable services and facilities at reasonable prices from alternative suppliers. Choice is a valued asset of a competitive market place.

In more formal terms, according to Bailey and Friedlander (1982, p. 1040):

> A perfectly contestable market has the following properties: (1) potential entrants are assumed to face the same set of productive techniques and market demands as those available to incumbent firms; (2) there are no legal restrictions on market entry or exit, and no special costs that must be borne by an entrant that do not fall on incumbents as well, that is, the technology may offer scale economies, but must not require sunk costs; and (3) incumbent firms can only change prices with a non-zero time lag (the price-sustainability assumption), but consumers respond to price differences with a shorter lag.

Intuitively, contestability can be viewed as providing conditions under which economies of scale, as such, pose no threat to market efficiency because they are not necessarily associated with Stiglerian relative barriers to entry. Economies of scale, if they reflect properties of production processes available to many firms, do not constitute a barrier to entry.

Perfectly contestable markets provide a benchmark for socially efficient market performance in naturally oligopolistic markets as Ramsey (inverse price elasticity of demand) prices will emerge. Formally, in the absence of income effects (for simplicity) for socially optimal subsidy-free prices we want prices (p) to maximize $CS(p) + R(p) > c$ (where CS and R refer to industry consumer surplus and profit). Thus, even in the natural monopoly case, *if* the market is perfectly contestable then it will automatically exhibit desirable welfare properties.

It is only the crucial feature of vulnerability to hit and run entry that perfectly competitive and contestable markets have in common. As long as there is the threat of entry when there are supernormal short-run profits, and exodus without cost when these profits are eliminated, contestable markets will force firms to operate in the P = MC pareto optimal manner.

Economists have even shown that in certain situations there can be a natural monopoly market (that is, subadditive costs) and yet there may be no set of prices sustainable against entry. When prices are not sustainable, barriers to entry need to be included within the regulatory set (see Haring, 1984, p. 23).

Contestability of Telecommunications

It has been argued that the national long-distance telecommunications market is contestable. Certainly the conclusions reached in studies by the DOJ (1983), Evans (1983), Evans and Heckman (1984), and Fuss and Waverman (in Fromm, 1981)[19], which indicate that telecommunications is not a natural monopoly[20], support the opinion that it is a contestable market. Nonetheless, in general the conditions required for perfect contestability or something close to it are demanding. Empirical research is desperately needed to reveal if the contestability conditions are commonplace and how close we are to them in certain industries.

A well established and generally accepted economic doctrine holds that public policy should encourage the conditions requisite to the contestability of a market. As Bailey and Friedlander (1983, p. 1042) put it, contestability has taught us, at a minimum, that "even though small may be beautiful, big isn't necessarily bad!" Furthermore, policymakers have begun to recognize that the boundary between the regulated and unregulated portions of an industry should be based on contestability (or subadditivity of costs) rather than on technological considerations. For instance, in recent years they have tended to favor continued regulation of those portions of the telecommunications business that have large sunk costs, such as local cables and wires, as regulatable. But segments of the industry that are perceived to be more inherently competitive are being opened to new entrants. For example, a series of decisions by the FCC has opened the terminal equipment markets, for both businesses and households, to vigorous competition while simultaneously trying to avoid inefficiencies in the utilization of the network.

A similar phenomena is occurring in the long-distance markets, or as Bailey and Friedlander (1982, pp. 1044–1045) have succinctly observed:

> With the advent of wireless transmissions systems, such as microwave services and satellites, policymakers have acted on their belief that the provi-

sions of network services has lost some of its technologically based attribute of either scale or scope or entry barriers. As a result, through a series of FCC decisions, new entrants are being permitted to compete in the broadest sense with network service offerings of the traditional telephone companies.

Relevant Elasticities of Demand and Supply

A study by Professors Landes and Posner (1981; also see the numerous "Comments" in 1982), emphasizes that market power depends on the price elasticity of demand and the fringe firm's elasticity of supply as well as on the market shares of each market participant. Thus, an examination of market shares alone can be misleading. In fact, a high market share is neither necessary nor sufficient for a firm to have market power (that is, raise prices above the competitive level). In general, greater market power comes from (1) a low (inelastic) market price elasticity of demand, (2) a low (inelastic) elasticity of supply on the part of the fringe firm, (3) and a high market share. We will examine each of these.

Market Direct Price Elasticity of Demand

In general, a high market direct price elasticity of demand implies that there are good substitutes for the service the firms sell and the existence of such substitutes limits the firms' market power. Conversely, the more inelastic are the price elasticities of demand, the greater is the opportunity for firms to exert their market power.

Previous national estimates of the coefficients of market direct price demand elasticity, which historically were generally of the order of −.2 to −.9 (inelastic) for toll service (Perl, 1983), were done before divestiture and therefore cannot be assumed relevant today.[22] Because numerous companies subscribe to an OIC and cite price as the main reason for their decision to start with a non–AT&T-C company for long-distance service, it is possible that that own price elasticity of demand is greater (more elastic) than previously estimated.

Fringe Firms' Price Supply Elasticity and Cross Elasticity of Supply

The higher the price elasticity of supply of competitive fringe firms, other things constant, the smaller the market power. A high supply elasticity means that a small price increase will lead to a large increase in the output of the competitive fringe. This supply expansion will be brought about by existing firms expanding and new firms entering the market. The more the expansion by fringe firms, the more an incumbent must reduce its output to maintain a given price. To capture at least partially the potential effects of supply elasticity, some suggest that the capacity or potential output of firms be considered by making market share calibrations of capacity.

DOJ Opinion on OIC Capability to Enter. The DOJ ("Comments", 1984, pp. 15–20) feels the OICs lack capacity to expand output easily and quickly. They believe that the addition of transmission capacity to serve new routes on a network involves considerable lead time. Moreover, they argue that extensive switching capacity and distribution costs must be sunk to establish a firm as a viable competitor in this market. DOJ further intones that to utilize microwave facilities, an OIC must buy or lease land. This often mandates rezoning, complying with local and federal environmental laws, and securing frequency allocations. To employ cable or optical fiber, a firm must secure rights-of-way. Further delays are occasioned by the need for management and logistical coordination to support the capacity expansion.

DOJ concedes that the capacity of the satellite carriers can be increased more easily and quickly than that of terrestrial-based carriers. However, the satellite carriers face at least two problems in augmenting capacity. First, to the extent that they need to supplement their networks with terrestrial expansion, they face the problems described earlier. Second, the extant and planned orbital allocations put limits on the amount of new capacity that can be introduced to companies with AT&T-C.

Furthermore, because the OICs' networks are considerably smaller and less developed than AT&T-C's, even a large percentage increase in their capacity represents only a small absolute enlargement in the total capacity of the market.[23] Even if MCI, the largest OIC, were to double its capacity, it would add only a few percentage points to the overall capacity of the market.

On the other hand, there is considerable evidence of substantial capacity expansion at least nationally by OICs. MCI and GTE-Sprint will spend about the same $1 billion for new capacity in 1985 as AT&T-C will, which is similar to what they each did in 1984. Some private bypassers are offering capacity for hire on the networks they control and manage. All told, private bypassers are expanding capacity by adding 35,000 microwave circuits in 1984 alone. Microwave transmission capacity doubled between 1974 and 1983 and OICs now account for 72 percent of all route-mile capacity. Satellite capacity is expected to grow nearly 200 percent from 1982 to 1987. By then, AT&T-C will own less than 11 percent of the total capacity. In 1978, OIC's revenues totaled $270 million; at the end of 1983, they totaled $3.7 billion. Facility providers (such as Fibretrak and Lightnet) are offering essentially turnkey services over 13,000 miles of network to over 50 cities to serve high volume customers. DTS licenses have been granted for service in over 200 cities to 68 firms.

Evidence in Georgia

While some econometric investigation of the relevant demand elasticities for the state of Georgia is available, we are unaware of any econometric estimates of supply elasticities. With present data sources and econometric techniques, econometric estimates of supply elasticities would be a herculean task. But tolerably accurate surrogates could be assessed with less effort. For instance, estimates of supply elasticities could be pieced together from such things as the capacity and market share statistics from the FCC's Form M. Estimates of demand elasticities can also be developed from the available surveys and other sources.

Demand Characteristics

In general, the more that demand is highly skewed and rapidly growing, everything else the same, the less likely it is that market power can be exercised. Thus, several characteristics of the demand for inter-LATA service promote competition, and these must be considered in assessing market power (see Kaserman, 1984). First, both residential and business demand distributions are highly skewed, or rather, a small percentage of the users accounts for a very large percentage of the traffic volume and revenues. This makes each carrier's demand highly vulnerable to competitive actions by competitors.

Since the relatively high-volume customers are thought to be very price sensitive, each firm will be forced to keep its price as competitive to protect itself against the potential loss of these valued users. Whether the loss is to competing carriers or to self supply, the result is the same—a powerful incentive to charge low prices. Moreover, these low prices cannot be limited to the class of high-volume users. Any attempts to reduce prices to the high-volume users while maintaining relatively higher prices to the low-volume customers (where such higher prices are not justified by real cost differences in providing service to these groups) will be unsuccessful because of the incentives and abilities to arbitrage. Resellers have already entered the arbitraging market to ensure that the benefits of active competition for the business of the high-volume customer will also apply to the low-volume customer.

In addition, the demand for inter-LATA telephone services has been growing dramatically. This surge of demand is likely to continue for some time and will provide a strong incentive for new firms to enter the market.

The emergence of new carriers will tend to force price downward as a larger number of carriers are competing for the users' business.

Statistical Evidence. There is supportive national data on the skewness of demand (*AT&T Outlook,* 1984). In the market, about 15 percent of the business customers (those high-volume customers who spend $50 or more per month on toll calls) generate about 90 percent of the interstate, long-distance business revenues. OICs provide service to about one-third of all those customers and to almost one-half of such customers located in areas OICs have chosen to serve. Similarly, high-volume residence customers with monthly bills of $25 or more represent only 10 percent of all households but generate 51 percent of interstate residence revenue. In the areas they have chosen to offer service, OICs now serve one out of five of these types of customers. OICs, of course, target the profitable high-volume customer segment for their sales efforts. In target areas where at least one competitor has chosen to enter, studies show that OICs provide service to more than half (56 percent) of the business customers who spend more than $150 a month on toll service. Those customers account for 75 percent of all business interstate revenue in OIC-served areas.

Regulatory Changes

In general, the more that a contestable nonnatural monopoly is deregulated, everything else the same, the less likely it is that market power can be retained or acquired. Because telecommunications has historically been a regulated industry, a comprehensive analysis of market power must take due note of important regulatory changes. Although we have emphasized the changes at the federal level, it is also important to keep in mind statewide changes. Similarly, it is important to remember that the public interest theory of regulation emphasizes that direct regulation can improve on competition only when the firms are : (1) natural monopolies so that a single firm can provide service more cheaply than several firms and eschews wasteful duplication; and (2) providing essential services, accessible to all, at reasonable prices. Whatever the merits of (2), the move toward deregulation in telecommunications has largely come about because of doubts as to the *current* applicability of (1).[24]

Three Sweeping Deregulatory Changes

Three sweeping changes have begun at the FCC which have facilitated the reduction of AT&T-C's previous market power and enhanced the growth of OIC competition. First, in a series of decisions, the FCC eliminated or moderated some or all of the impedimenta that had previously prevented new carriers from entering the toll market, and removed constraints on pricing and service offerings by these new carriers. These profound alterations of the regulatory milieu was a first step to unbridling competitive market forces.

Also, while this is subject to dispute, the deregulation reduced incentive to engage in either predatory pricing or *pricing without regard to cost,* as the DOJ claimed in its antitrust case against AT&T (see Evans, 1983, pp. 55–56, 213).[25] Although the naive version of the proposition (that a regulated firm has a greater incentive to engage in economically predatory cross-subsidization) is flawed, unless one assumes that AT&T-C has other incentives besides profit maximizing, the reduction in political predation (barrier-to-entry investment) without deregulation is defensible. It should be noted, however, that when one looks at all the factors that are needed for economic predation to be profitable—asymmetrical information, uncertainty, risk aversion, reputation—we can see why the most reasonable conclusion that one can draw from economic theory and from trial records is that both regulated and nonregulated firms have considerable profit incentives *not* to engage in economic predation. As to pricing without regard to cost, this is simply an impossible practice in the long run for a surviving profit-maximizing firm in a (perfectly) competitive or contestable market.

Second, the divestiture of the BOCs from AT&T effectively eliminated the possibility that AT&T-C could exclude rivals from the inter-LATA market by denying access to the local loop entirely or even price-discriminatorily. This modification of the industry's economic structure guarantees that franchised market power in one sector of the industry cannot be employed for leverage in another sector.

Third, the equal access provision of the MFJ, concomitant with the unequal carrier access charges that will remain in effect until equal access is achieved, represents the final step in placing all carriers (both actual and potential) on parity.

Evidence in Georgia

We are at this point unaware of any steps taken or contemplated by the Georgia Public Service Commission which would change the general tenor of our conclusions based on the national picture.

Q-Ratio and Profits Measures of Market Power

In general, the higher the *q*-ratio and the higher the long-run economic profitability, everything else the same, the more likely it is that market power is in force. Thus, circumstantial evidence that is often used to determine the existence and magnitude of market power is the q-ratio and (economic, not accounting) profitability. Although our discussion is limited at this stage to national considerations, it would be instructive for future researchers to get at least an order of magnitude of the state-specific figures these indicate. However, we do not believe this would be feasible for the q-ratio and only slightly less difficult for profitability.

Q-Ratio

Some recent economic literature, especially Lindenberg and Ross, (1981) and Salinger, (1984) has popularized Tobin's q-ratio as a measure of the extent, distribution, and history of monopoly rents and quasi-rents.

Salinger demonstrates why Tobin's q is a better measure of monopoly profits than indices of single-period profitability. The q-ratio measures long-run monopoly power, is less subject to measurement error, contains an adjustment for risk.

In a pioneering work, James Tobin introduced the variable q, defined as the ratio of the market value of a firm's securities (common and preferred stock as well as debt) to the replacement cost of all of its assets, into macroeconomics to examine the causal relationship between q and investment. Later on, Lindeberg and Ross, and Salinger applied it to microeconomics with outstanding success. In brief, Lindenberg and Ross used Tobin's q to measure monopoly power in the U.S. economy. Theoretically, q should be near 1 for competitive firms and greater than 1 for firms with monopoly power.

If q were greater than 1 in a competitive milieu, the entry of new firms would push it back towards 1. In contrast, a monopolist can earn monopoly profits (or quasi-rents) by barriers to entry, including possessing unique factors of production, sales organizations, or products, and strong patent protection or by inadequate regulation. The market value of the securities of these monopolists exceeds their replacement costs as q-ratios are persistently greater than 1.

In their study of the (average) q-ratios of 257 firms for the period 1960–77, Lindenberg and Ross found AT&T's (average) q-ratio to be 1.09, well below that of a firm expected to be receiving monopoly returns. The (average) q-ratios for all firms ranged from .45 (Cone Mills Corporation) to 8.53 (Avon Products). Other interesting q-ratios included General Telephone and Electronics 1.32, IBM 4.21, and RCA 1.67. Salinger found a mean average q-ratio for the 252 nonregulated utility manufacturing firms to be 1.18 for the end of 1976.

Profitability

Not only is the *q*-ratio low for AT&T, but it has been demonstrated that profits have also been low. For instance, MacAvoy and Robinson, (1983, p. 37, footnote 152) find "an examination of prices established by regulatory commissions during the 1970s reveals rates of return insufficient to cover Bell's cost of capital. In fact, AT&T's return was consistently below the return for a composite of Standard & Poor's 400 industrials."

The profitability of some OICs also flies in the face of conventional wisdom. For instance, the five-year return on equity for MCI, the largest OIC, exceeds the overall industry average by 15 percentage points and is over twice that of the AT&T-C. If AT&T-C does possess market power, it should be the firm exhibiting above-average profitability, *ceteris paribus*.

The maintained hypothesis in economics and by the DOJ ("Guidelines", 1984, p. S-8) is that if *long-run economic profitability* over substantial periods of time significantly exceeds that of firms comparable in risk and capital intensity, this is circumstantial evidence of possible noncompetitive performance. However, because economic profits, in contrast to accounting profits, consider all the implicit opportunity costs and revenues, and not merely the explicit costs and revenues, they are exceedingly difficult to determine. A number of adjustments must be made to accounting profits to get even an order of magnitude on economic profits (see Fisher and McGowan, 1983). But it may be worth the effort. For as Schmalensee (1982, p. 1808) states: "Depending on the fact of the case at hand, data on profitability or on patterns of conduct may be more informative than are market shares."

Notes

1. According to the *DOJ* ("Guidelines", 1984, p. S-7) market power is the "ability of one or more firms profitably to maintain prices above competitive levels for a significant period of time".

2. "Asymmetric regulation" is the name Richard Schmalensee (1984) uses for the FCC's present policy of subjecting AT&T-C to more stringent regulation than its rivals.

3. In the technical language of economists, a local (global) subadditive cost function is a necessary and sufficient condition for a local (global) natural monopoly. Formally, subadditivity exists if the costs of joint production are less than the costs of separate production for any scale of output or combination of outputs. It should be noted that multiproduct economies of scale are neither a necessary nor a sufficient condition for the existence of subadditivity. As we shall see later, economies of scope are a necessary condition and economies of scope and product specific economies of scale are a sufficient condition for subadditivity (see Evans, 1983, Chaps, 6, 10).

4. Fuss and Waverman (in Fromm, 1981) have also suggested, on the basis of data for Bell Canada from 1952–1975, that there was a real question whether the Canadian telecommunications industry is a natural monopoly. On the other hand, Phillips (1982) argues that the pre-divestiture evidence in the United States and Canada exhibits subadditivity of costs.

5. The main ideas of contestability are presented in Baumol, Panzar, and Willig, (1982). See also Waterson, (1984, pp. 73, 80). The most comprehensive attack to date is that of Shepherd, (1984).

6. Remember *market power* gives a firm or cooperating group of firms a degree of discretion in controlling the price, the quantity, and the nature of the products offered for sale. See any standard industrial organization textbook such as F. M. Scherer (1980, pp. 59–64).

7. The principles underlying most of the factors discussed here can be found in any standard price theory text such as Kamerschen and Valentine, (1981), or industrial organization textbooks such as Clarkson and Miller (1982).

8. You will recall that an economic market is a collection of products in a specified geographical area that, if offered by a single seller, could increase profitably the price above competitive levels for a sustained period of time. Hence, the necessary and at least superficially sufficient conditions for a market are (1) that the products involved be considered substitutable by the buyers and sellers and (2) be comparably priced.

9. While the recent economics literature has emphasized exit barriers as an important separate element in assessing likely market performance, the previous literature assumption that exit barriers are comparable to or lower than entry barriers remains as yet empirically unrefuted. We have therefore emphasized entry over exit conditions, yet we are cognizant of the latter's potential significance.

10. For the Stigler-Bain distinction see, e.g., H. Craig Peterson, (1985, pp. 29–32).

11. That is, companies' antitrust immunity in joining together to bring lawsuits or otherwise petition the government is void when the actions are considered "sham" attempts to further the companies' own interests.

12. Even if a firm has a scale economies (or subadditive) costs, this does not necessarily mean that the industry is a natural monopoly and therefore single-firm production is more efficient than multi-firm production and competition should be prohibited. Several other assumptions—indentical utilization of technology, necessity of common asset ownership, regulatory efficiency, and costlessness—are also necessary to complete this connection. Although single-product scale economies imply a single-product natural monopoly, (1) product-specific economies in both products do not necessarily imply a multiproduct natural monopoly; and (2) a multiproduct natural monopoly may not necessarily have product-specific scale economies in either product. Thus, product-specific economies of scale are of limited value in the determining if a natural monopoly exists or not (see Evans, 1983, p. 130–133).

13. See American Telephone and Telegraph, 1983 and Western Electric, 1983.

14. In only one of the seven telephone regions will this not occur. In the Northeast Region, which encompasses New York and New England, AT&T and its competitors will be served from the same switches in the same LATAs.

15. The BOCs state that identical per-call access would be impossible, even if a uniform access configuration were imposed in each LATA, because of "normal variations of electrical characteristics of facility components and installation line-up variations. . . ."

16. The BOCs are not responsible, of course, for correcting any quality deficiencies that may result from an interexchange carrier's own facilities.

17. A Point-of-Presence (POP) is the physical location of interconnection between a local teleco and an IC. The latter compensate the former for the cost of the transport of both originating and terminating calls, based on time and distance.

18. That is, if the sunk costs of establishing telephone service are substantial, the company that established service initially would realize a first mover advantage. Because of these advantages, there is a tendency to build too soon and capture the market, although concomitant innovative behavior may offset this. Thus, the FCC's delay in allowing Datran to build a nationwide digital network for transmitting digital information, requiring a large lump-sum investment, may have been socially optimal.

19. They conclude in a recent and comprehensive study that Bell Canada did not have: (1) aggregate scale economies; and (2) a natural monopoly over local, toll, and private line services.

20. This includes the belief (Evans, 1983, p. 34) that local exchange monopolies may never have been "natural."

21. Direct (cross) demand price elasticity measures the percentage change in the quantity demanded of a product in response to the percentage change in the price of that (another) product.

22. However, the Wharton (1983) study on access charges used average long-distance (local service) direct market price elasticities from Taylor (1980) of -1.31 (-0.10) for residential and -1.17 (-0.09) for business. Thus, the long-distance coefficients are elastic and the local service coefficients are inelastic.

23. AT&T-C has about $9 billion worth of interstate net plant, whereas the combined net plant of the OICs is estimated at $5–6 billion.

24. Anyone who questions the statement that modern civilization would be unthinkable without rapid communication would be well advised to recall reports of the incident where Andrew Jackson, lacking rapid communications, fought the British at New Orleans after peace had been reached in the war of 1812 (see Phillips, 1984, p. 617).

25. It was filed on November 20, 1974 charging AT&T has used its dominant position to suppress competition and enhance its monopoly power. The 1982 MFJ represents the final settlement of it.

References

AT&T Bell Laboratories, *Engineering and Operations in the Bell System, 2nd ed.,* (Murray Hill, N.J., 1984).

Bailey, Elizabeth E., and Friedlander, Ann F., "Market Structure and Multiproduct

Industries," *Journal of Economic Literature,* Vol. 20 (September, 1982), pp. 1024–1048.

Baumol, William J., Panzar, John C., and Willig, Robert D., *Contestable Markets and the Theory of Industry Structure.* San Diego: Harcourt Brace Jovanovich, 1982.

Besen, Stanley M., Woodbury, John R., "Regulation, Deregulation, and Antitrust in the Telecommunications Industry," Vol. 28, *The Antitrust Bulletin* (Spring 1983), pp. 39–68.

Brennan, Timothy J., "Mistaken Elasticities and Misleading Rates," *Harvard Law Review,* Vol. 95, (June, 1982), pp. 1849–1856.

Bonbright, James C., *Principles of Public Utility Rates.* New York: Columbia University Press, 1961.

Danielsen, Albert L., and Kamerschen, David R., Eds., *Current Issues in Public-Utility Economics: Essays in Honor of James C. Bonbright,* Lexington, Mass: D.C. Heath, Lexington Books, 1983.

Dixit, Avinash K., "Recent Developments in Oligopoly Theory," *American Economic Review, Papers and Proceedings,* Vol. 72, (May, 1982), pp. 12–17.

Evans, David S., Ed., *Breaking Up Bell: Essays on Industrial Organization and Regulation,* New York: North-Holland, 1983.

Evans, David S., and Heckman, James J., "A Test for Subadditivity of the Cost Function with an Application to the Bell System," *American Economic Review,* Vol. 74, (September, 1984), pp. 615–623.

Federal Communications Commission, *Statistics of Communications Common Carriers.* Issued annually since 1939.

Fisher Franklin M., "Diagnosing Monopoly," *Quarterly Review of Economics and Business,* Vol. 19, (Summer, 1979), pp. 7–33.

Fisher, Franklin M., and McGowan, John J., "On the Misuse of Accounting Rates of Return to Infer Monopoly Profits," *American Economic Review,* Vol. 73, (March, 1983), pp. 82–97.

Fuss, Melvyn, and Waverman, Leonard, "Regulation and the Multiproduct Firm: The Case of Telecommunications in Canada," in Gary Fromm, Ed., *Studies in Public Regulation.* (Cambridge: MIT Press, 1981), Chap. 6, pp. 277–313; also see the "Comment" in *idem* by Ronald Braeutigam, pp. 314–320, and "Comment", also in *idem* by Bridger M. Mitchell, pp. 321–327.

Garfinkel, Lawrence, "Interexchange Telecommunications Markets in Transition," *Public Utilities Fortnightly,* Vol. 112, No. 2, (July 21, 1983), pp. 26–33.

Joskow, Paul L., and Schmalensee, Richard, *Markets for Power: An Analysis of Electric Utility Deregulation.* Cambridge: MIT Press, 1983.

Kahn, Alfred E., "The Road to More Intelligent Telephone Pricing," *Yale Journal of Regulation,* Vol. 1, (No. , 1984), pp. 139–157.

Kahn, Alfred E., "The Uneasy Marriage of Regulation and Competition," *Telematics,* September 1984.

Kaplow, Louis, "The Accuracy of Traditional Market Power Analysis and A Direct Adjustment Alternative," *Harvard Law Review,* Vol. 95, (June, 1982), pp. 1817–1848.

Kaserman, David L., Direct Testimony, Corporation Commission of Kansas, Docket No. 127, (October, 1984), pp. 1–29.

Katz, Michael L., and Willig, Robert D., "The Case for Freeing AT&T," *Regulation,* Vol. 7, (July/August 1983), pp. 43–49.

Kelley, Daniel, "Deregulation After Divestiture: The Effect of the AT&T Settlement on Competition," OPP Working Paper #8, April 1982.

Landes, William M., and Posner, Richard A., "Market Power in Antitrust Cases," *Harvard Law Review,* Vol. 94, (March 1981), pp. 937–996.

Lexitel Corporation, "Comments," FCC Docket No. 83-1147, (April 2, 1984), p. 21.

Lindenberg, Eric B., and Ross, Stephen, "Tobin's Q Ratio and Industrial Organization," *Journal of Business,* Vol. 54, (January, 1981), pp. 1–32.

MacAvoy, Paul W., and Robinson, Kenneth, "Winning By Losing: The AT&T Settlement and its Impact on Telecommunications" *Yale Journal of Regulation,* Vol. 1, (No. 1, 1983), pp. 1–42.

Microwave Services International, Inc., *Private Microwave Status Reports,* (Denville, NJ: MSI Telecom Consultants Group, June 1, 1984).

Ordover, Januz, Sykes, Alan, and Willig, Robert, "Herfindahl Concentrations, Rivalry, and Mergers," *Harvard Law Review,* Vol. 95, (June, 1982), pp. 1857–1874.

Owen, Bruce M., and Braeutigam, Ronald, *The Regulation Game.* Cambridge: Ballinger, 1978.

Perl, Lewis J., *Residential Demand for Telephone Service,* 1983. White Plains, New York: National Economic Research Associates, December 16, 1983.

Phillips, Almarin, "The Impossibility of Competition in Telecommunications: Public Policy Gone Awry," in *Regulatory Reform and Public Utilities,* Michael Crew (ed.), 1982.

Phillips, Almarin, "The Pseudoeconomics of Access Charges," in *Boundaries Between Competition and Economic Regulation,* J. Rhoads Foster *et al.* (eds.), 1983.

Phillips, Charles W., Jr. *The Regulation of Public Utilities: Theory and Practice.*Arlington: Public Utilities Reports, Inc., 1984.

Robertson, Reuben B., and Wiley, Richard E., Chairman, *The Breakup of AT&T: Opportunities, Prospects, Challenges.* Law & Business, Inc./Harcourt Brace Jovanovich, 1982.

Salinger, Michael A., "Tobin's Q, Unionization, and the Concentration-Profits Relationship," *Rand Journal of Economics*, Vol. 15, (Summer, 1984), pp. 159–170.

Salomon Brothers, Inc., "MCI Communications Corporation—Can We Talk?," (August 1, 1984), pp. 1–20.

F.M. Scherer *Industrial Market Structure and Economic Performance, 2nd ed.* Chicago: Rand McNally, 1980.

Schmalensee, Richard, "Another Look at Market Power," *Harvard Law Review,* Vol. 95, (June, 1982), pp. 1789-1816.

Schmalensee, Richard, "Statement of Richard Schmalensee," Attachment 4 to Comments of AT&T in CC Docket No. 83-1147, April 2, 1984.

Shepherd, William G., "Contestability vs Competition," *American Economic Review,* Vol. 74, (September, 1984), pp. 572–587.

Shooshan, Harry M., III, Ed., *Disconnecting Bell: The Impact of the AT&T Divestiture.* New York: Pergamon Press, 1984.

Sullivan, Brian, "Equal Access—Who Benefits, Who Pays?," in *Changing Patterns in Regulation, Markets, and Technology: The Effect on Public Utility Pricing,* Harry Trebing (ed.), 1984.

Taylor, Lester D., *Telecommunications Demand: A Survey and Critique,* Cambridge: Ballinger, 1980.

U.S. Department of Justice, "Comments," FCC Docket No. 83-1147, (April 2, 1984).

U.S. Department of Justice," Merger Guidelines issued by the U.S. Department of Justice, June 4, 1984, and Accompanying Policy Statement," *Antitrust and Trade Regulation Report,* Special Supplement, (Washington, D.C.: The Bureau of National Affairs, Inc., No. 1169, June 14, 1984, pp. S1–S16).

Varian, Hal R., *Microeconomic Analysis, 2nd ed.* New York: W.W. Norton, 1984.

Waterson, Michael, *Economic Theory of the Industry.* Cambridge: Cambridge University Press, 1984.

Wharton Economic Forecasting Associates, *Impact of the FCC Access Charge Plan on the U.S. Economy,* (November, 1983), pp. 1–19.

Legal References

American Tobacco Co. et al. v. *United States,* 328 U.S. 781 (1946).

Berkey Photo v. *Eastman Kodak Co.,* 603 F.2d 263 (2 Cir. 1979), cert, denied, 444 U.S. 1093 (1980).

Eastern Railroad Presidents Conference v. *Noerr Motor Freight Inc.* 365 U.S. 127 (1961).

United Mine Workers v. *Pennington U.S.* 381 U.S., 657, 670 (1965).

United States v. *American Telephone and Telegraph Company,* 552 F.Supp. 131, 227 (D.D.C. 1982) *aff'd. sub. nom. United States v. Maryland,* 103 Sup. Ct. 1240 (1983) and *United States v. Western Electric Co.,* 569 F.Supp. 1057, 1062–1064 (D.D.C. 1983).

United States v. *E.I. duPont de Nemours & Company,* 351 U.S. 377 (1956).

18
Interexchange Issues

Robert E. Fortenberry
ATT Communications

If the ultimate goal of the telecommunications industry today—full and free competition—were fully achieved, this chapter might not be necessary. If competition fully ruled the marketplace, all we would be concerned about is meeting customer needs and doing it better. And that would mean our major interests would be marketing and technology.

The fact is, however, that there are a lot of so-called interexchange issues to be concerned about. The fact is that we have come a long way in changing this industry in recent years. And the fact remains that we still have a long way to go to achieve the ultimate goal of today's public policy—full and complete competition in the interexchange or long-distance marketplace.

A Look Back

Without technological advances such as microwave, transistors, and telecommunications satellites, without microchips and, even more basic, without the novel idea that common, ordinary sand could be used to produce vital information-age products—from solar cells to fiber-optic cable—there would be no reason to discuss the competitive long-distance marketplace at all.

In my view, the marriage of technology to customer needs is the single most powerful force driving our industry today. Technology has helped us achieve universal telephone service, and I believe it will help us achieve the industry's new goal: universal information service—the ability to provide any customer with any kind of voice, data, or image capability, any place, any time, and with maximum convenience and economy.

The political, economic, and structural issues we are struggling with today are no more than temporary barriers to the age of universal information service. And if there is one lesson to be learned from the past, it is that government policy cannot, in the long run, resist the imperatives of technology and customer demands. Of course, AT&T is pleased that it played the major role in these technological developments—developments that made competition a reality.

Competition started with telephones, then went to private-line networks, and then to switched long-distance services. It has meant that we have had to create a new dictionary for the industry for terms such as *access charges, other common carriers, divestiture, equal access, allocation, points of presence,* and so on. Of course, this means more acronyms. We now have BOCs, R-BOCs, and LECs. POPs now joins POTS in the industry lexicon, as well. These are terms generated as the industry moves toward competition under the watchful eyes of adjudicators, legislators, and regulators. Hearings continue in courtrooms, in Congress, and before regulators—both state and federal.

Although there are many competitors in the long-distance business today, the marketplace is still not allowed to be fully competitive. Today we have regulated competition. The term *regulated competition* has become the industry's oxymoron. I am confident that one day it will be a term of the past—and competition will be played out fully under consumer control in the marketplace, not in hearing rooms.

When do we allow the competitive marketplace to take over? At what point does regulation let go? How fast and to what extent should cost-based pricing be introduced, and when? How should long-distance subsidies to local service be phased out? How do we protect universal service? All of these are legitimate and critical issues in our industry today—issues that competitors, regulators, judges, legislators, and industry experts are working furiously to resolve.

Do Not Forget Why We Are Doing This

My concern is that in this process, we do not get caught up in the process itself—hearings, testimony, legal proceedings, and so on—and forget why all of this is being done.

The fact is that all these changes are happening because it has been decided that the American consumer can now benefit more from telecommunications advances by obtaining services in a competitive marketplace. Just as the American consumer shops competitively for automobiles and appliances and, more recently, for banking services and air travel, so should he or she be allowed to shop for telecommunications services. All this is being done for the American consumer, ultimately. So, I think that we should look at today's long-distance issues from the consumer's point of view. Is today's marketplace serving that consumer as it should?

The State of Competition

To answer that question, I would like first to take a quick look at the state of competition in long distance. As I said, we have regulated competition.

However, the regulation generally applies only to AT&T in the interstate marketplace and does not apply to its more than 300 competitors.

In various state jurisdictions, the rules are different. Some regulate only AT&T, some regulate all carriers equally and pervasively, and in Georgia and a growing number of states, regulators are gradually beginning to allow the competitive marketplace to take over. About half the states in which interexchange competition exists now have some form of reduced regulation for all long-distance carriers. This generally means the ability to change prices without full regulatory proceedings. In Georgia, carriers are allowed to change rates at any time with fourteen days' public notice, so long as the new rates are below levels the GPSC set. Other states, such as Mississippi, use a banded-rate approach. Rates may change on short notice anywhere between established maximum and minimum levels. In Nevada and Virginia, it is "no holds barred." In Oklahoma, a recent order removed rate-base regulation, and an Idaho order puts private lines completely in a competitive mode.

What these commissions are acknowledging is that there is a competitive long-distance world and that competition brings in the best regulator—the consumer. In the South, regulators in Virginia, South Carolina, North Carolina, Georgia, Mississippi, Tennessee, and Louisiana have now reduced the surrogate role they played for the consumer in the past days of a one-supplier marketplace. It is appropriate that they are turning the decisions and choices offered by an increasingly competitive long-distance marketplace back to the consumer. Subsequent rate reductions in many of those states would tend to show that without regulatory control, the pressure is indeed on the competitors to reduce their prices. So it is indeed reasonable for an attitude of regulatory forebearance to take hold—that is, stepping back and letting competitive pressures take hold. And although there may still be a way to go to achieve this forebearance, there has been considerable progress.

Customer Service

Other issues facing the industry directly affect customer service:

Computer Inquiry II

AT&T is particularly interested in alleviating the Federal Communications Commission's Computer Inquiry II (CI-II) rules. These pre-divestiture rules were developed to prevent the local-exchange side of the old Bell System business from subsidizing the competitive parts.

Judge Greene saw to it that local exchange and noncompetitive parts of the Bell System are no longer with AT&T. Therefore, CI-II rules are an answer looking for a question. The problem is that customers are affected. ATT Communications and ATT Information Systems must make separate presentations when a customer is interested in hearing about a total communications package from AT&T. We have competitors, some larger than we are, who can offer similar systems but are not required by law to put the customer through separate sales pitches.

In the customer equipment area, CI-II rules were slightly relaxed earlier this year, but many more decisions remain. In an equivalent of Hollywood's *Star Wars II* or *Return of the Jedi,* the FCC has now created what might be called "Return of the CI," or CI-III, to look at other issues. It also intends to examine why enhanced services have not come to the market in the manner expected from CI-II.

Service

Service, in general, has survived changes to a competitive marketplace, divestiture, and a number of regulatory decisions. Customers have fared well in some cases and not so well in others.

Basic long-distance service is as good as ever. AT&T initially had some private-line and WATS/800 provisioning problems as a result of divestiture, but those have been resolved. There is, indeed, more choice in the marketplace—at least for a large number of customers. Also, customers can now begin to designate any carrier choosing to participate in equal access by dialing one-plus. In the four southern states with which I am involved, approximately two-thirds of the Bell company's access lines have equal access.

On the downside, customers have had to get used to getting bills as fat as sweepstakes mailers. They also are not getting some new services as fast as they should. AT&T PRO℠ America and AT&T's Software Defined Network—all new service offerings—are being held up at the FCC for formal regulatory proceedings, proceedings that are not required of all companies. Many state commissions, including Georgia's, have had the foresight to make AT&T PRO offerings available to customers without undue regulatory delay. Each is a service that can provide customers money-saving opportunities.

Customers also have been faced with constantly changing rules in today's form of regulated competition. Just recently, customers found that they would be arbitrarily allocated to a primary long-distance company if they did not choose one in the carrier-selection process. Our customer service centers were flooded with calls the Monday after that Friday FCC decision. The Atlanta center handled some 19,000 calls instead of the usual 11,000. Many customers called our repair centers after hours, just to be sure that they could talk to someone. The Jacksonville, Florida, center handled more than 200 calls an hour more than average. Many people were angry at the arbitrary nature of the decision; many were confused; others, thankfully, were just calling to be sure they could stay with AT&T.

I will not go into all the ins and outs of allocation. My point here is that regulated competition is too often leaving customer considerations on the back burner. The sooner decisions are left to customers in a freely competitive marketplace, the better.

Price

I have saved the most critical issue for last. It is also probably the issue that continues to need the most attention—price.

It is no secret that for years and years, long-distance rates have been set artificially high to help subsidize the cost of local service. It is no secret that the subsidies continue under a new name today: carrier access charges. AT&T and, to a lesser extent, its competitors pay substantial dollars to local telephone companies for access that is priced far over cost. As a result, all long-distance customers are paying too much for long-distance service today—whether they are using AT&T or some other company. In fact, these access expenses are about 70 percent of my company's cost of doing business—and if you include other costs such as billing, the total cost approaches 75 to 80 percent.

The short-term fear is that if local costs are returned to the local rate structure, universal service will suffer. Customers will not be able to stay on the public network. The long-term fear, howelver, is that without moves to get prices for all telecommunications services closer to the cost of providing the service, universal service will suffer far more down the line. Why?

You have, no doubt, heard the term *bypass* (another word for the modern telecommunications dictionary). Basically, it is a fallout in a competitive world of continuing pricing practices used in the one-supplier marketplace of bygone days. Wonderful new technology is accessible and relatively inexpensive. Large customers know this and will build their own networks to avoid high long-distance bills. And when they build their own systems, they are often going further—they are also avoiding use of the local telephone company for exchange services.

It has long been a fact in our business that 10 to 15 percent of the largest business customers account for 90 percent of the long-distance revenue. It is not surprising that they also account for the greater share of subsidies resulting from artificially high long-distance rates. The question, then, becomes: Who pays when these big customers abandon the public network?

Studies and other evidence are piling up. Local phone companies, regulators, industry associations, NARUC, legislators, long-distance companies, and other experts now recognize the magnitude of this long-term threat to universal service. That is why the FCC's $1 customer line charge was finally implemented; that is why the $6 business charge was approved; and that is why some state commissions (most recently Alabama) have begun adopting some sort of customer line charge or similar cost-based pricing plan. These efforts mark a start. Market-based pricing, in which consumers drive prices toward cost, is a must in a competitive world. It is in the ultimate best interest of all consumers.

The transition has begun too slowly and must pick up steam to prevent further erosion of the major customer base the public network now has. Decisions made today definitely affect telecommunications prices in the next decade, and the sooner the market is allowed to control itself, the better off all consumers will be. The sooner we get to cost-based pricing, the lower long-distance bills will be ten years from now. And the sooner we get to cost-based pricing, the lower local service bills will be tomorrow.

Considered another way, the price of telecommunications services in the 1990s and beyond can never be as low as it could be if proper pricing decisions are made today. The sooner those decisions are made, the lower those prices will be.

Discounts

I would also like to address another access-related matter—the discounts AT&T's competitors receive on connections to local companies. Looked at from the customer's point of view, it is an issue of fairness. It is difficult for my customer in Vidalia to see the logic of his being forced to pay more for long distance so that a fellow Georgian in Atlanta can pay less.

The FCC still allows AT&T's competitors a 55 percent discount on interstate access payments to local companies until equal access conversions have been made. Most southern states are far below that, and many are reducing discounts for intrastate access well before equal access has been achieved.

The Tennessee Public Service Commission said it would allow no discount, stating that there is no proof of a real difference in cost of the connections. On the other hand, the commission also cited the inequity of asking a rural customer to subsidize fellow citizens in metropolitan areas.

Recently, four of AT&T's competitors have asked the FCC to extend the discount beyond equal access. (Not all companies chose to go along with that petition.) From comments reported after the petition was filed, it appears that many people inside the FCC are becoming more outspoken on attempts by competitors to use the regulatory process to protect profits. Nevertheless, the rules of the day call for comments by all parties, and a regulatory proceeding is under way to investigate this latest petition for special treatment.

Summary

Many more issues of the day could demand time; I have tried, in this chapter, to highlight the major ones. I think it is important that we in the industry, along with public policymakers, keep in mind that the original reason to move toward competition in telecommunications was to take full advantage of the opportunities new technology offered consumers.

At this point, however, we still have not reached the goal of good, old-fashioned American marketplace competition, in which the consumer serves as the ultimate regulator. We are locked somewhere between regulation and competition. As a result, there is often confusion. New services are often stifled; various competitors have opportunities to take advantage of regulatory processes; and the still-regulated price structure remains out of synchronization with the dictates of today's (and tomorrow's) competitive marketplace.

Do not get me wrong—I think we have made progress, and I am truly pleased to say that regulators, legislators, and industry people in the southeast are at the leading edge of moving toward competitive thinking in the best interests of consumers. Our charge in the future is to continue that progress—and to make haste to reap the fruits of a freely competitive marketplace that will best serve everyone in the information age.

19
Depreciation Reserve Assessment

Terence Robinson
GTE Service Corporation

Domestic telephone companies continue to face increasing levels of competition from both unregulated and lightly regulated carriers. It has become apparent that existing depreciation practices and procedures—formulated decades ago, in times of limited competition and a relatively stable state-of-the-art technology—now need to be supplemented with a broadened capital recovery perspective. This capital recovery perspective must recognize both the industry's significant move toward greater levels of competition within traditionally protected franchise areas and the explosive rate of technological innovation that is driving down the real cost of productive capacity.

Moving from a depreciation perspective and toward a capital recovery perspective requires that we fully embrace the theory and concepts inherent within economic selection theory; revenue requirement regulation; the time value of money; and accounting theory. A capital recovery system satisfying the constraints imposed by each of these disciplines would define the theoretically correct pace of capital recovery as follows: *A method of book depreciation that, when combined with the appropriate level of tax depreciation and the investment tax credit, will maintain a balance between the capital invested in an asset and that asset's economic value.*

In the remainder of this chapter, we will develop an economic model that enables us to measure the value to society of selected technologies at various points in their expected life span.

The Competitive Assumption

The industry economic value model developed in this chapter embraces the assumptions associated with the traditional economic model of perfect competition. The competitive model lends itself to structured economic analysis because the behavior of the market is constrained by objective forces that are subject to rational analysis. Also, the economic model of perfectly competitive markets is the reference framework used by economists and regulators to assess the appropriateness of proposed regulatory economic policy.

The competitive model is used to develop regulatory policy designed to provide the benefits of competition to the subscribers of a regulated company's services. For the consumer, the primary benefit of competition is prices that are lower than would be charged in an environment in which competition is restrained.

We feel compelled to build the foundation of our model on the assumptions of perfect competition, because the telecommunications market is becoming increasingly more competitive. To shift the capital recovery process to accommodate these conditions, it is necessary to identify the behavior that is appropriate in such markets. In addition, the resulting prices from such analyses should reflect what regulation should be attempting to achieve in those markets that are not fully competitive.

The Impact of Technology

If technological innovation is reducing the real costs of production, prices, in real terms, will be reduced over time. Nominal prices may be either increasing or decreasing, depending on the rate of inflation. The rate at which real prices are reduced will mirror the rate at which technology is reducing real costs, so that a "normal" real return is retained. Therefore, at equilibrium in the presence of technological innovation, each supplier will face a downward trend in real prices that will provide a "normal" return over the life of a technology. Figure 19–1 illustrates market revenue and cost functions over the life span of a given technology.

For purposes of this illustration, assume that there is no inflation and that there are no income taxes. A producer that acquires productive capacity employing the best available technology will incur ongoing costs to operate the asset so long as it remains in service. Revenues will diminish over time, as competition reduces the market price. The price is then reduced by competition, because technological innovation is reducing the real cost of new productive capacity. The economic life of the technology will end when prices fall to the point at which the costs of production employing a given technology exceed the revenue available from the best available technology.

In figure 19–2, we overlay the cost function from an existing older technology on the market revenue and cost functions for the best available technology. The point at which the older technology cost curve (old cost) intersects the market revenue curve represents the end of the older technology's economic life. In this example, the point of intersection occurs fourteen years following introduction of the new technology.

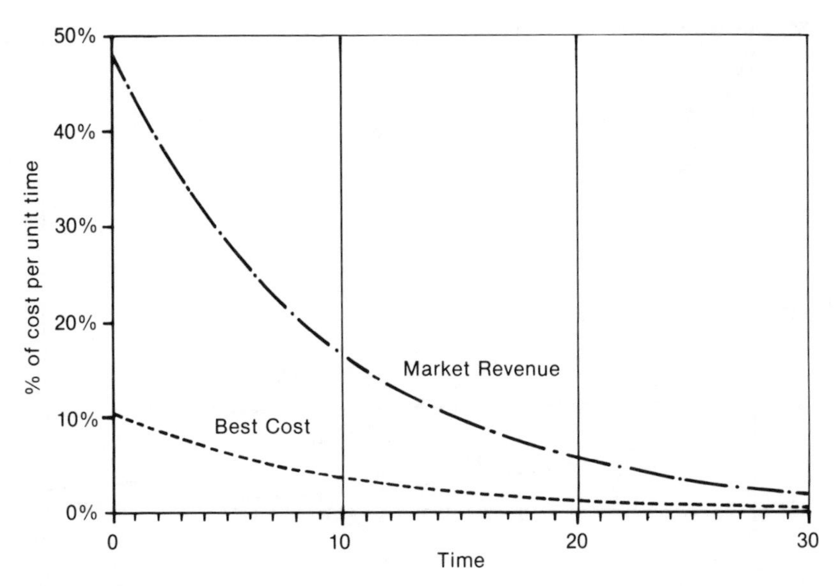

Figure 19–1. Market Revenue and Cost Functions: Over the Life Span of a Given Technology

Concept of Measuring Societal Value

The value to society of a given technology may be approximated by projecting the net present value of the technology's future net cash flow expectations. In the context of our current generic example, this would represent the shaded area in figure 19–3. The present value of this area represents the expected value to society of the new technology measured in year 0.

In the case of the reserve assessment analysis, we are interested in measuring a technology's remaining value at a future time (1990). For purposes of illustration, assume that we are interested in determining our example technology's remaining value in year 10. The value is represented by the shaded area in figure 19–4. If we determine the present value of this area, discounted back to year 0, we have an approximation of the technology's remaining value. The relationship of the remaining value to the total original value gives us a proxy for a technology's required reserve.

Given the economics associated with the best available technology, we can measure the remaining economic value of an older technology by following the same procedure. This is illustrated in figure 19–5. Note that the cost functions for both technologies begin at approximately the same level. However, the value of the older technology falls quickly because of the new technology's enormous impact on the slope of the market revenue curve. By

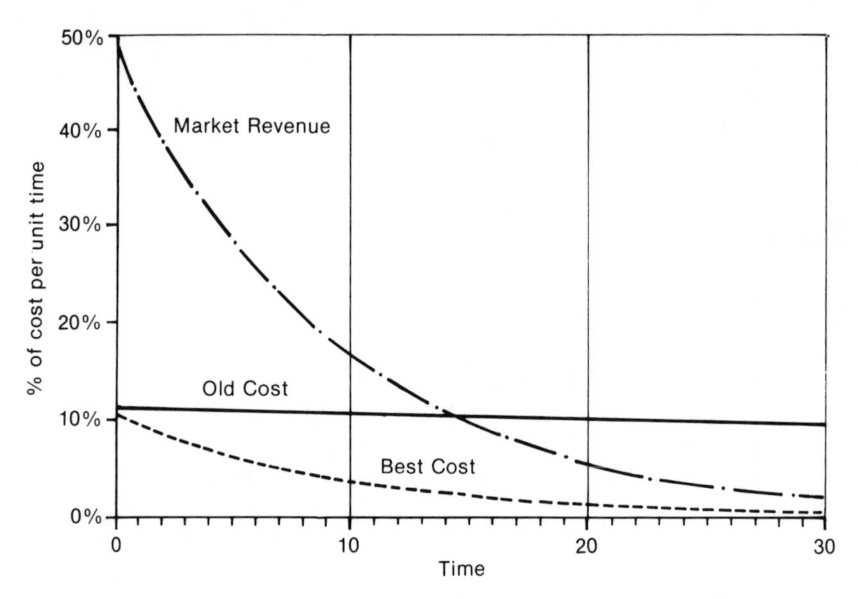

Figure 19–2. Market Revenue and Cost Functions: Old Technology versus Best Available Technology

year 10, very little economic value remains in the older technology's investment capacity.

The concept of value outlined in this section tracks closely with our definition of capital recovery. In our definition, we stated that the appropriate pace of capital recovery must recognize impacts of income taxes as well as book depreciation and must maintain a balance between economic value and net book. The pattern of economic value discussed in this section represents rate base or net book value—that is, original cost less book depreciation reserves and deferred tax and ITC tax reserves.

This distinction of value is important because, under the economic depreciation model, the book depreciation of an asset may be very low (even negative) during the early years of life (depending on the key parameter values discussed in the following section). This occurs because of tax benefits during the first part of an asset's life. For example, a new digital central office costing $1 million and having an economic life of fifteen years will generate approximately $.24 million in tax benefits during its first year of service.

We should also note, at this point, that various tax reform proposals now being considered by the U.S. Treasury Department will have a major effect on our value-based depreciation reserve ratios, to the extent that they tamper with the existing investment tax credit and ACRS depreciation sched-

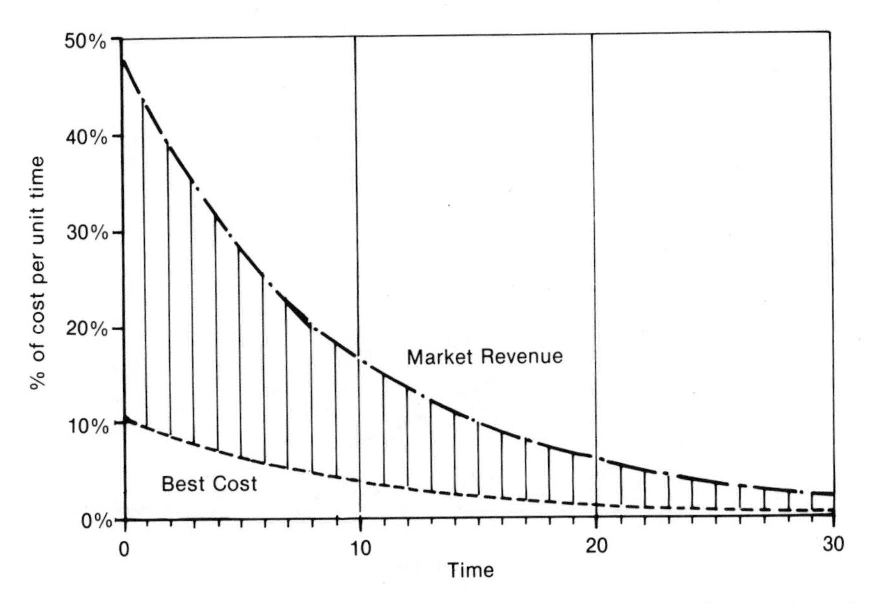

Figure 19–3. Market Revenue and Cost Functions: Net Present Value of Future of Given Technology Net Cash Flow Expectations

ules. (See the GTE Service Corporation white paper on economic value depreciation for a comprehensive review of this subject matter.)

Applying the Model

The preceding section of this chapter outlined the philosophy and conceptual underpinnings of our industry economic value model. In this section, we will briefly describe the actual application of the mathematics inherent in the economic value depreciation model and the application of those concepts in the industry economic value model. The basic equation underlying the concept of economic value depreciation is as follows:

$$V_0 = \sum_{n=1}^{L} \frac{[R_1(1-p)n-1-O_1(1+g)n-1][1+in(1-t)+t(\mathrm{TD}_n+\mathrm{ITC})]}{\{1+[(1+K_e)(1+i)-1]c+(1-t)[(1+K_d)(1+i)-1](1-c)\}^n} \tag{19.1}$$

with constraints

$$R_1(1-p)^{L-1} \geq O_1(1+g)^{L-1}$$
$$R_1(1-p)^{L} < O_1(1+g)^{L}$$

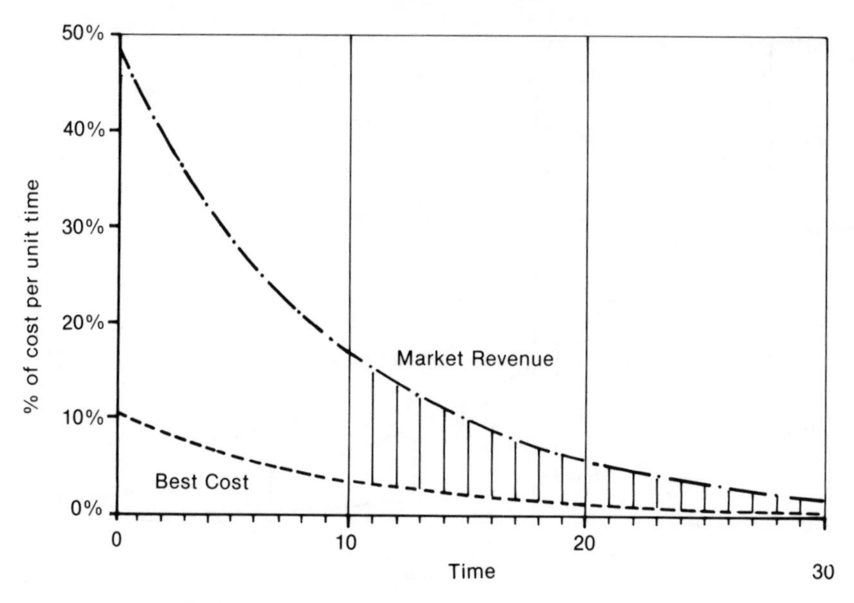

Figure 19–4. Market Revenue and Cost Functions: Technology's Remaining Value at a Future Time

where V_0 = purchase price of a unit of productive capacity
R_1 = revenue in period one
O_1 = operating cost in period one
p = percent decline in real replacement cost due to technological innovation
g = percent increase in the real cost of operating expenses over the economic life
L = economic life of the asset
i = a measure of inflation
t = income tax rate
ITC = investment tax credit
K_e = real cost of equity
K_d = real cost of debt
c = perent equity to total capital
TD_n = each year's tax depreciation.

Equation 19.1 represents a condition of equilibrium in which the present value of all future net cash flows, including the market rate of return in each period, must equal the original investment cost (V_0). The constraints require that the aggregate revenue exceed aggregate cost in each year of the invest-

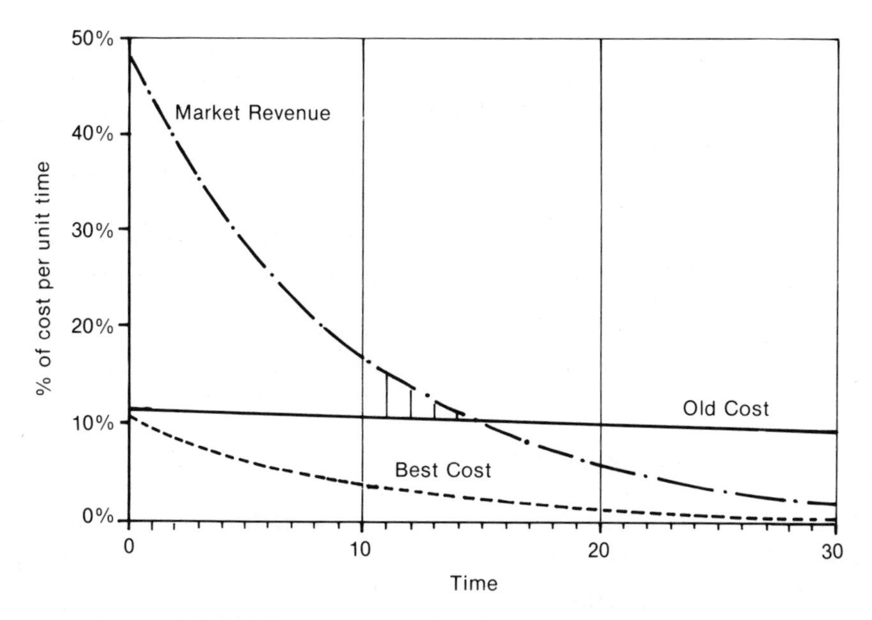

Figure 19–5. Market Revenue and Cost Functions: Remaining Economic Value of an Older Technology

ment's economic life and that in the year following the end of the economic life, cost must exceed revenue. The denominator is simply the firm's (or market's) after-tax cost of capital, which is being used as the appropriate discount rate. The numerator is the net cash flows expected during each period of the investment's economic life, including income tax benefits. We can estimate values for each of the required variables with the exception of R_1 and L, where R_1 is the revenue required in period one and L is the economic life of the asset. Given values for the other variables, there is only one combination of values for R_1 and L that satisfies the two model constraints.

We use equation 19.1 as a model to generate the market revenue curve for each network business segment. We apply the equation using the best available technology input parameters. Our assumptions and values for each variable will be presented later in the chapter. By applying the model over a relatively long span of time and adjusting V_0 in each period to reflect the changing price of a productive unit of capacity, given our inflation and technological innovation assumptions, we are able to generate a series of $Ri1$ values that represent a technology's life cycle market revenue or price curve.

We also use equation 19.1 as a model to generate avoidable cost curves for each technology participating in a particular network business segment.

The model is applied in the same manner as described earlier for the market revenue curve. The result is a generic cost curve for each technology participating in the market, which recognizes the unique operating characteristics of each technology.

Once we establish the market revenue and avoidable cost curves for each competing technology, we compute net present values for each technology. We follow a value determination procedure much the same as outlined earlier in our generic example. The results of our total and remaining value determinations leave us with industry reserve ratio requirements for each competing technology (including depreciation and tax reserves). We then back out the tax reserves to illustrate book depreciation reserve requirements.

Network Business Segment Results

Our study segments the network into two primary businesses—switching and transmission. Within the switching business, we study three competing technologies: SPC digital, SPC analog, and electromechanical. The transmission business is segregated into interoffice, exchange, and circuit equipment groupings. We study three competing technologies within the interoffice grouping: fiber optics, radio, and copper cable.

For the current planning horizon, the exchange grouping is expected to be predominantly copper cable, and the circuit equipment grouping is assumed to follow the evolution of the switching network.

Switching Business

For the immediate planning horizon, the best available technology in switching is assumed to be SPC digital. Competing with the SPC digital technology are the SPC analog and electromechanical technologies. The assumptions made for various parameters within the model, as discussed earlier, are outlined in table 19–1. Based on these assumptions, our model generates the market revenue and cost functions presented in figure 19–6.

Figures 19–7 and 19–8 illustrate the economic value remaining beyond 1990 for each competing technology (shaded area). Note that our best available technology—digital—has a relatively large portion of its total life span value remaining in 1990. The electromechanical technology has a significant value remaining beyond 1990, whereas the SPC analog technology has only a very insignificant value remaining. Our results indicate significant value remaining in the electromechanical technology primarily because of the relatively low cost of reusing displaced electromechanical switches, compared with the cost of purchasing new SPC digital equipment. The digital conversion plans are displacing a large amount of electromechanical capacity in

Table 19–1
Switching Business Model Parameter Assumptions

Item	*SPC Digital*	*SPC Analog*	*Electromechanical*
Common			
Inflation	5.78%	5.78%	5.78%
Discount rate	12%	12%	12%
ITC rate	8%	8%	8%
Tax rate	46%	46%	46%
ACRS life	5 yrs.	5 yrs.	5 yrs.
Unique			
g (expense change)	0	0	.3%
Operating costs	15%	16.9%	23.3%
p (technology)	(5%)	(2%)	(1%)
Cost/unit	100%	140%	70%

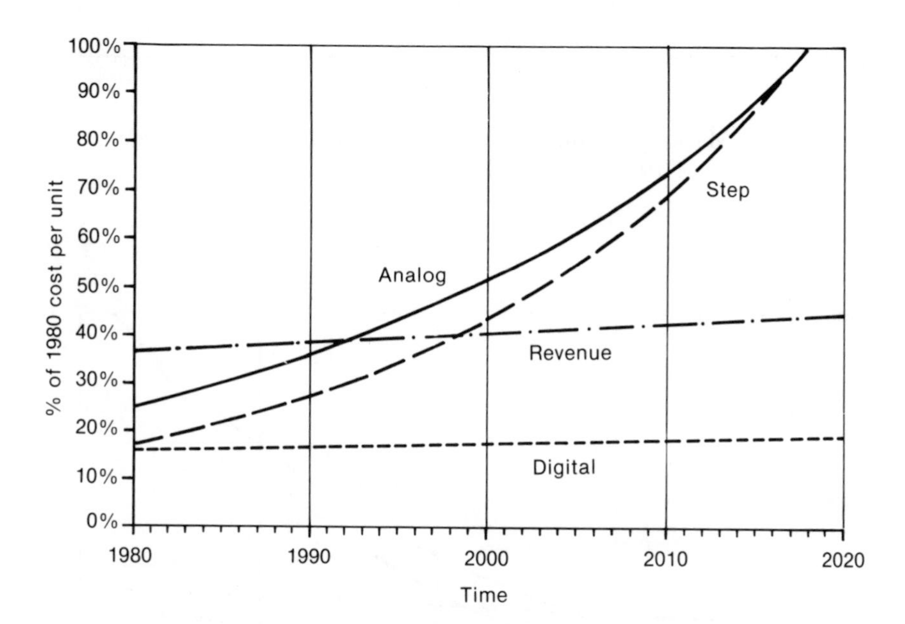

Figure 19–6. Market Revenue and Cost Functions: SPC Digital Technology Assumed to Be Best Available Technology

our primary markets, a technology that will continue viable in our secondary markets for at least the short run. The study does not suggest that new electromechanical offices will be installed, but rather intends to establish the technology's viability over time, given its cost characteristics. Our study indicates that the analog technology has very little value remaining. Much of

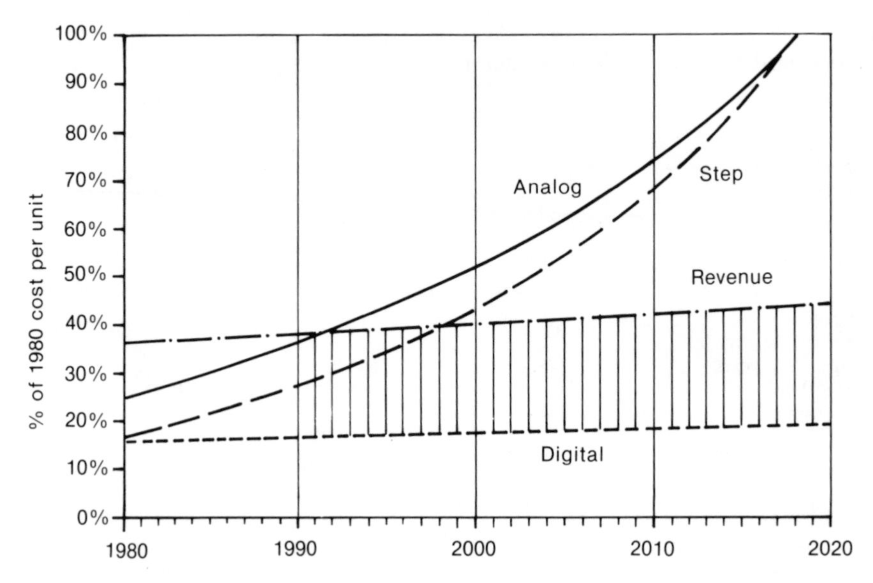

Figure 19-7. Market Revenue and Cost Functions: Economic Value Remaining beyond 1990, SPC Analog

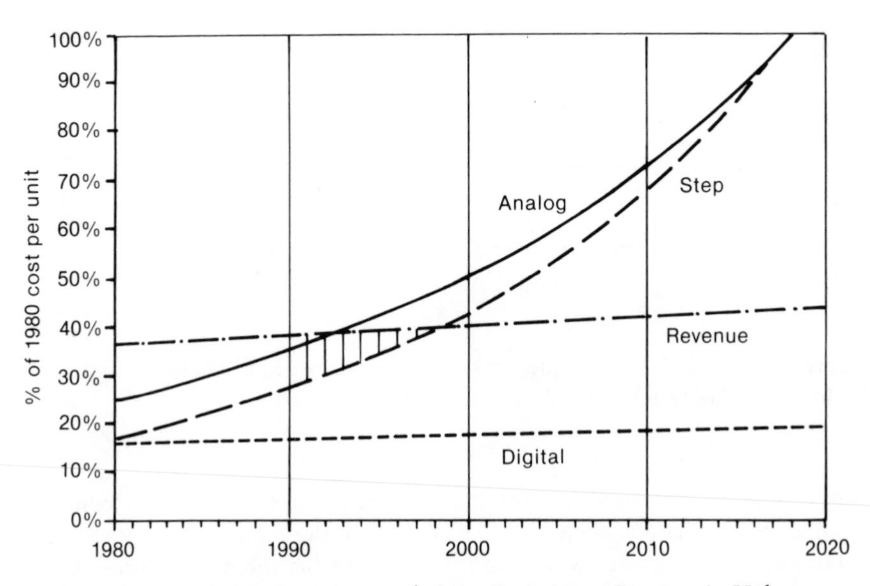

Figure 19–8. Market Revenue and Cost Functions: Economic Value Remaining beyond 1990, Electromechanical Technology

our existing analog capacity is in primary market areas scheduled for early conversion to SPC digital.

Figure 19–9 illustrates the net market residual revenue beyond 1990 for each competing technology.

The figures make it appear that the SPC digital technology has an infinite net revenue producing life. This is misleading in terms of a technology's total value to society for two reasons. First, in establishing value, we are interested in the present value of all future cash flows expected from that technology. In this chapter, we have been using a market discount rate of 12 percent. As figure 19–10 illustrates, the present value of the SPC digital technology's net cash flow over a seventy-year period is only a fraction of an absolute, undiscounted level of those same net revenues.

Second, a new technology will occur at some future time. When this new technology becomes viable in the marketplace, it will establish a new market revenue price function. Figure 19–11 reproduces the SPC digital revenue and cost functions, presented previously (old revenue and old cost). We now assume that a major technological innovation occurs in the year 2000 (for example, fiberoptical switch). In the year 2004 the new technology becomes the best available viable technology and establishes a new market revenue price function (new revenue). Now, instead of appearing to have an infinite

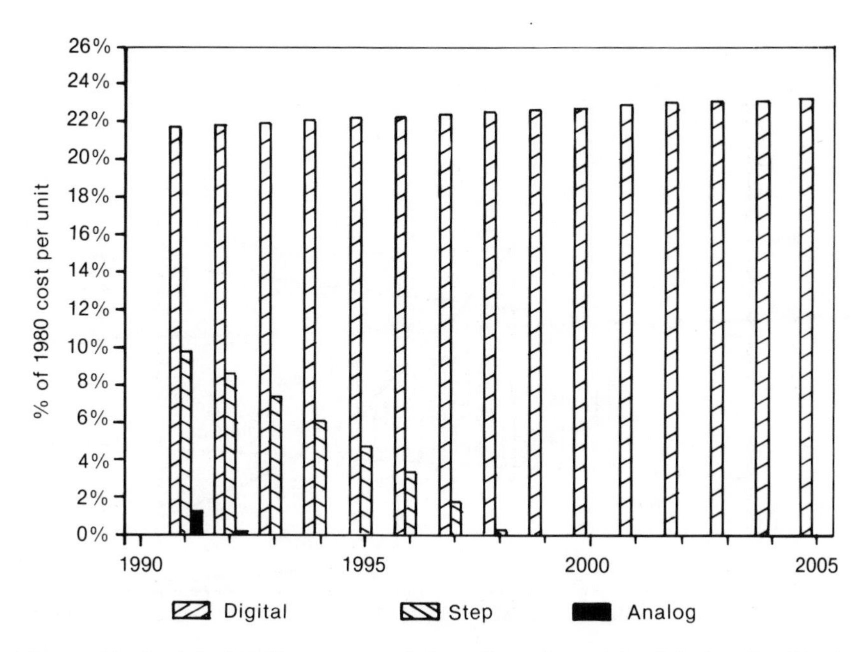

Figure 19–9. Market Revenue and Cost Functions: Net Market Residual Revenue beyond 1990 for Each Competing Technology

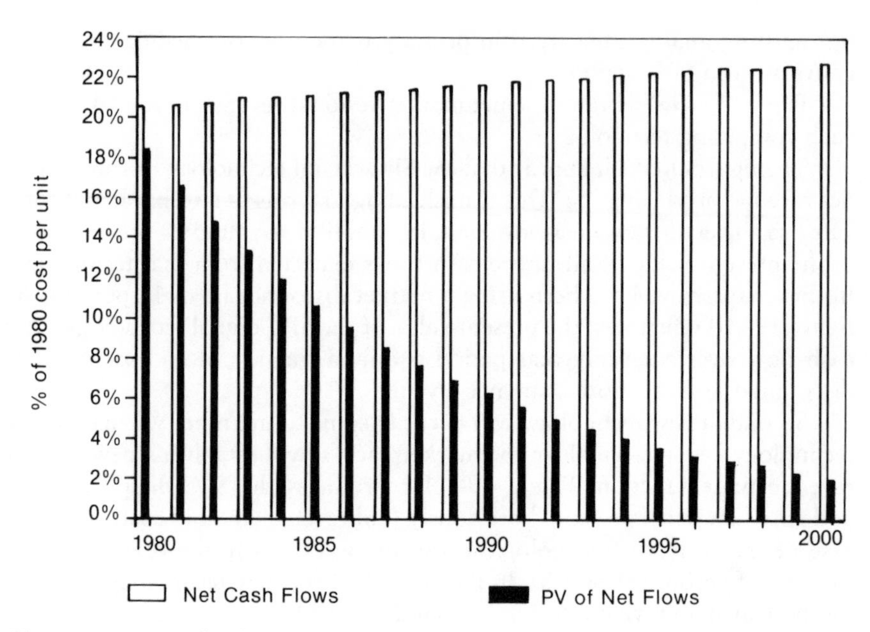

Figure 19–10. Absolute vs. Discounted Cash Flows: Digital Switching

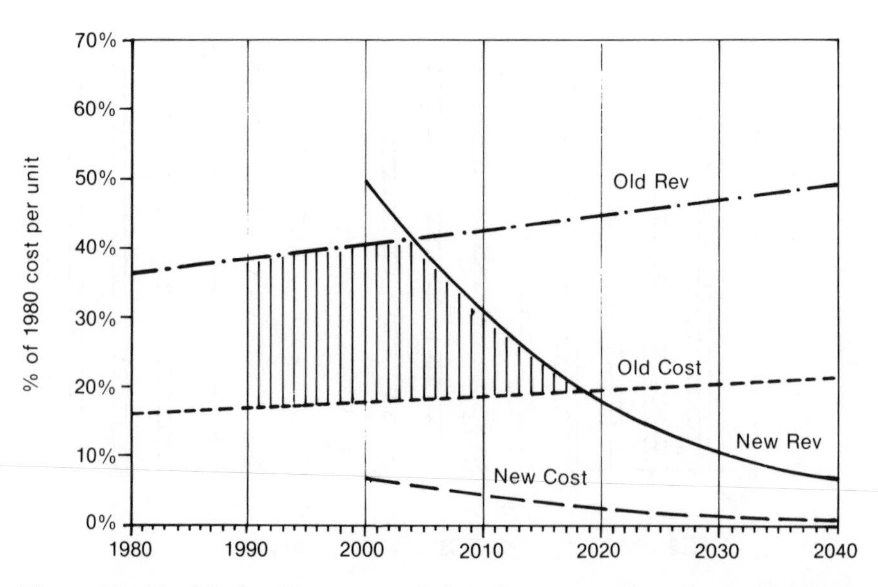

Figure 19–11. Market Revenue and Cost Functions: Introduction of New Technology 2000

net revenue producing life, SPC digital technology's life is abruptly cut off in the year 2019, which is the point at which the avoidable cost of using digital exceeds the market price constraint of the best available technology (fiberoptical switch). This, of course, has very little impact on our value assessment, because the present value of the digital technology's cash flows beyond approximately the year 2000 are pretty much insignificant (figure 19–10).

Application of the economic value concepts discussed earlier (in the context of our model) results in a required reserve ratio for 1990. We present two sets of reserve ratios in table 19–2. The column labeled "Economic Book Depreciation Reserve Ratios" represents the book depreciation reserve ratio requirement under the assumptions of economic value depreciation. The "Combined Depreciation" column presents the total reserve requirement, including the ITC and deferred tax reserves. This second set of ratios corresponds to ratebase or net book value (for example, in 1990 the value of SPC digital investment is 100 percent less 41.5 percent, or 58.5 percent).

Transmission Business

In the interoffice grouping, the best available technology is assumed to be either fiber optics or radio, depending on the specific application. In both cases, the best available technology is assessed against the competing copper cable technology. In exchange business, copper cable is assumed to be the dominant technology in the near term, and it is set at a point of decline (that is, beyond steady state) equal to 50 percent. Circuit equipment parameters are assumed to follow those of the SPC digital technology. The specific assumptions made for various parameters within the model are outlined in table 19–3. Based on these assumptions, our model generates the market revenue and cost curves presented in figures 19–12 and 19–13. Figure 19–12 illustrates the results when fiber optics is the best available technology, and figure 19–13 illustrates radio as the best available technology.

Figures 19–14 and 19–15 illustrate the economic value remaining beyond 1990 for the fiber and radio technologies, respectively. Note that in

Table 19–2
Required Reserve Ratios for Switching Business 1990

Switching Business	*Economic Book Depreciation Reserve Ratios*	*Combined Depreciation Reserve and Tax Reserve Ratios*
Electromechanical	82.7%	91.8%
SPC analog	97.9%	99.3%
SPC digital	7.2%	41.5%

markets such as the interoffice business, where we assume that fiber and radio are viable technologies, copper cable has no economic value remaining.

Figures 19–16 and 19–17 illustrate the net market revenue residual beyond 1990 for each technology competing in the interoffice business. Note the rapidly declining net revenues for the fiber technology. This is because of the extremely high (15 percent) annual rate of technological innovation and its impact on the price of a unit of productive capacity. In the case of radio, the decline in net market revenues over time is much more gradual. This is a more moderate slope because the rate of technological innovation (6 percent) only marginally offsets the rate of inflation (5.78 percent).

By applying the present value concepts in the same fashion as we did for the switching business, we derive the required reserve ratios for 1990 presented in table 19–4. Again we present two sets of ratios.

Industry Composite Reserve Ratio for Network Plant

The final step necessary to apply the modeled economic reserve ratios to the USTA exchange carriers is to composite the generic industry ratios developed here, based on each exchange carrier's projected 1990 mix of investment in each competing technology. We have done this for the USTA member exchange carriers, and we have arrived at a book depreciation reserve requirement of approximately 46 percent and a total reserve requirement (that is, including tax reserves) of approximately 66 percent.

Analysis and Recommendations

In reviewing the industry results presented in the earlier sections of this chapter, we find that we must be reaching network reserve ratios in the 40 to 50 percent range by the end of 1990. This does not seem unreasonable, given that such a large portion of our network plant is copper cable serving relatively low-growth markets and is therefore essentially operating in a steady state of decline. Because of this, we set the copper cable depreciation reserve ratio equal to 50 percent, which exceeds the steady-state results of 33 percent (assuming that copper is still the best available technology for exchange purposes).

If we assume that a major innovation occurs within the fiber technology within our planning horizon, making it desirable to push fiber further out into the network, our results change significantly. This is true because the copper cable technology will lost its economic value much faster than we have assumed in the study. For example, if such an innovation in fiber technology occurs and copper cable technology's required depreciation reserve ratio moves from 50 percent to 60 percent, the industry composite network reserve ratio would increase from 46 percent to 51 percent. If the copper cable depreciation reserve requirement moves to 70 percent, the industry composite network reserve would increase to 56 percent.

The reserve ratios presented in this chapter should be accepted as the best economically derived estimates available at this time of what the competitive marketplace demands. These reserve ratio objectives should be reassessed as new industry planning data become available.

Table 19–3
Transmission Business Model Parameter Assumptions

Item	*Fiber*	*Radio*	*Copper*	*Circuit*
Common				
Inflation	5.78%	5.78%	5.78%	5.78%
Discount rate	12%	12%	12%	12%
ITC rate	8%	8%	8%	8%
Tax rate	46%	46%	46%	46%
Unique				
ACRS life	10 yrs.	5 yrs.	15 yrs.	5 yrs.
g (expense change)	0	1%	2%	0
Operating costs	10%	16.5%	20%	15%
p (technology)	(15%)	(6%)	(1.5%)	(5%)
Cost/unit	100%	65%	80%	100%

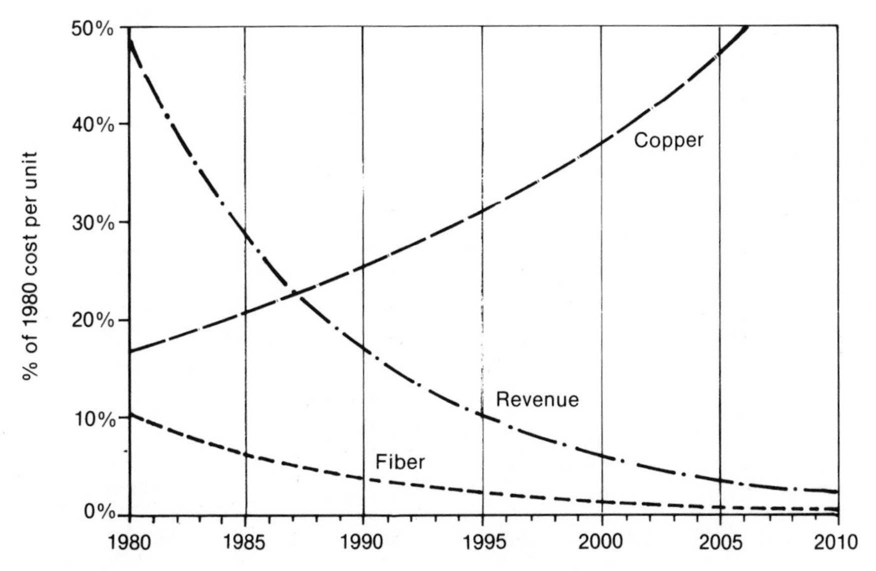

Figure 19–12. Market Revenue and Cost Functions: Fiber Optics Is the Best Available Technology

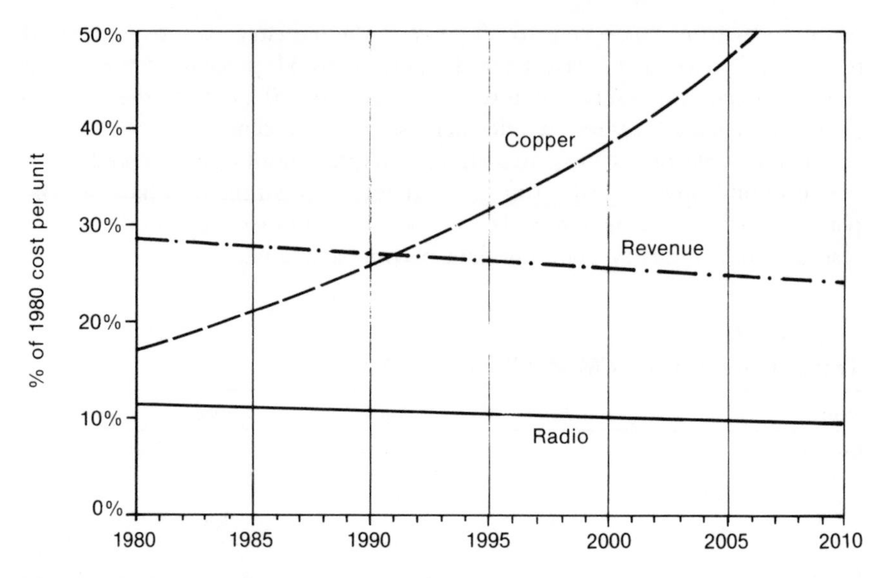

Figure 19–13. Market Revenue and Cost Functions: Radio Is the Best Available Technology

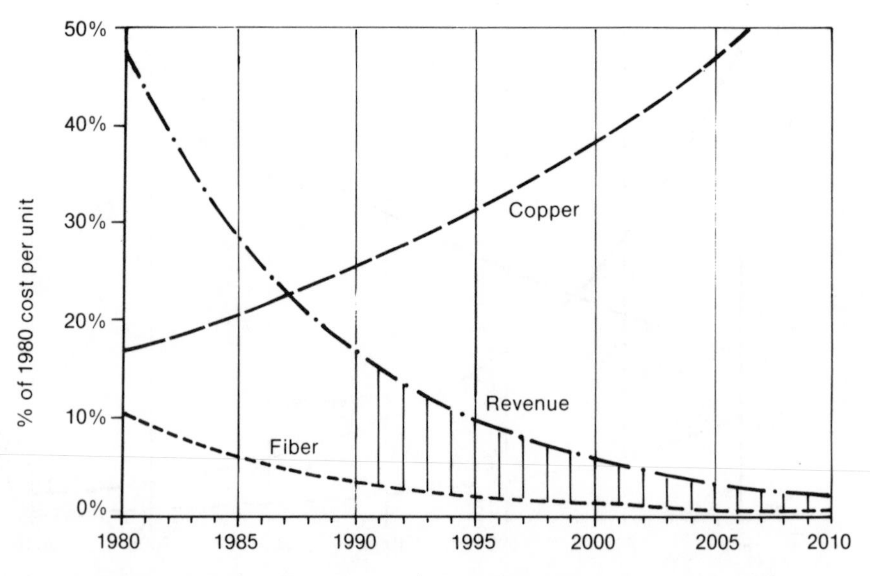

Figure 19–14. Market Revenue and Cost Functions: Economic Value Remaining beyond 1990 for the Fiber Technology

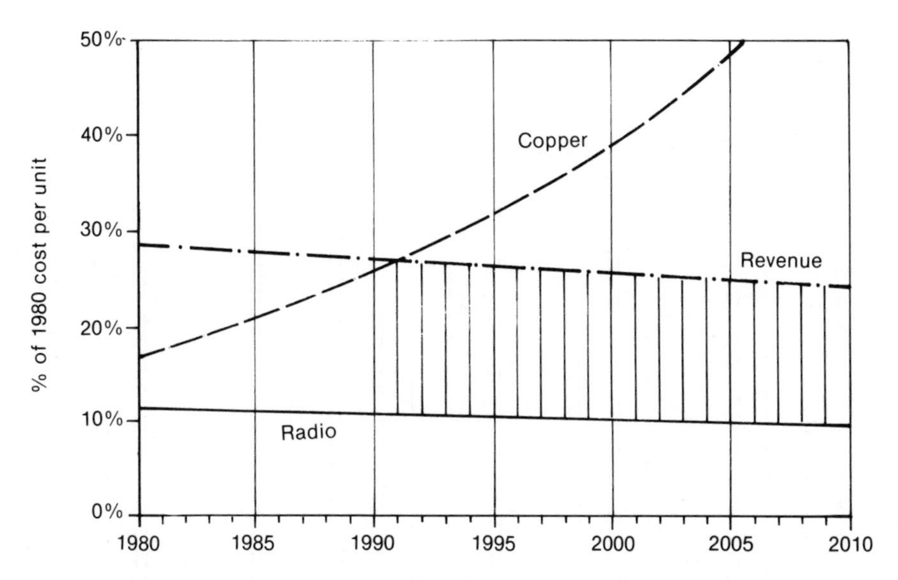

Figure 19–15. Market Revenue and Cost Functions: Economic Value Remaining beyond 1990 for the Radio Technology

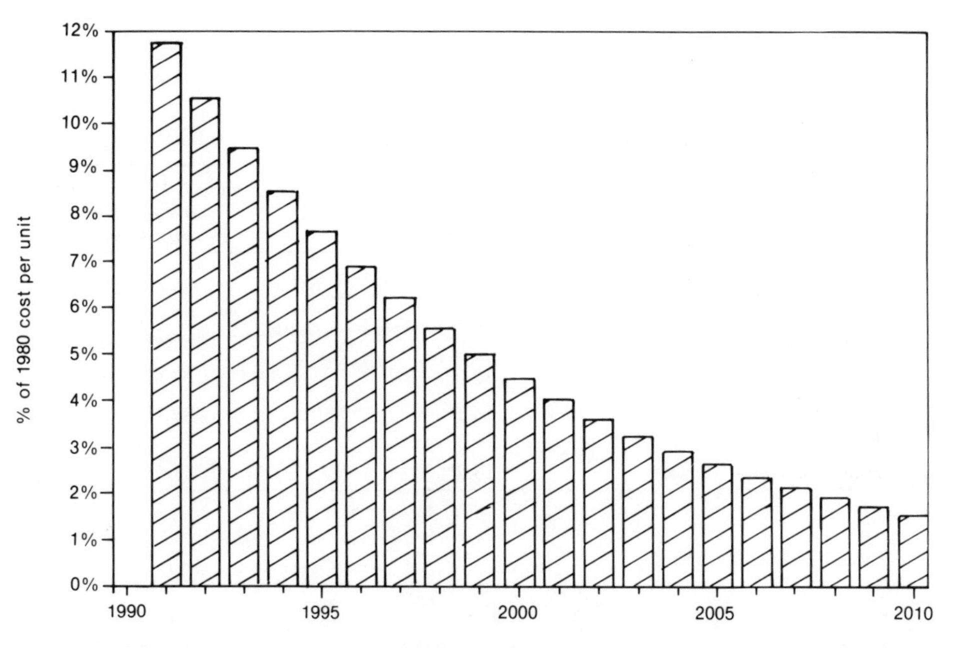

Figure 19–16. Net Technology Market Revenue: Transmission Business—Fiber

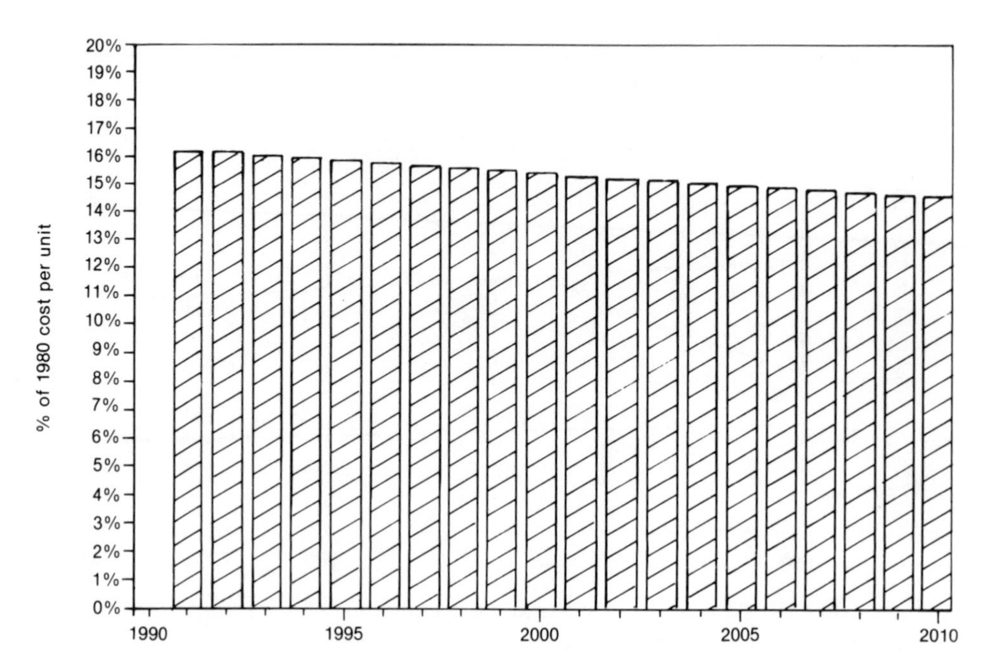

Figure 19–17. Net Technology Market Revenue: Transmission Business—Radio

Table 19–4
Required Ratios for Transmission Business 1990

Transmission Business	*Economic Book Depreciation Reserve Ratios*	*Combined Depreciation Reserve and Tax Reserve Ratios*
Interoffice facilities		
Radio	61.6%	82.9%
Fiber optics	34.2%	37.6%
Copper cable	50.0%	67.4%
Exchange facilities	50.0%	67.4%
Circuit equipment—analog	56.1%	80.1%
Circuit equipment—digital	13.3%	51.8%

20
A Comparative Analysis of Two Difficult Market Transitions: Telecommunications and Natural Gas

William F. Hederman, Jr.
Interstate Natural Gas Association of America

Since the late 1950s, the telecommunications industry has undergone a remarkable transformation. The monopoly on interstate telephone service earlier held by American Telephone and Telegraph has been eroded by the rapid growth of competing firms. AT&T's hold on the bulk of the nation's local telephone service, through its ownership of the Bell Operating Companies, has been broken by the divestiture of these companies that took place on January 1, 1984. The dependence of subscribers on their local telephone companies as sole suppliers of telephones and related equipment has been replaced by a highly competitive market offering choices scarcely imagined a few years ago. In short, the industry no longer fits the traditional public utility mold.

The interstate natural gas pipeline industry is also facing increased competitive pressures. Not only is it threatened by the loss of customers to oil, but gas-on-gas competition is a growing phenomenon as well.

Both industries are confronted with the problem of maintaining their public utility obligations, especially in supplying service at just and reasonable prices to residential customers in a marketplace where the ground rules are rapidly changing. They are thus confronted with the potential for encountering the worst of both worlds of competition and regulation. Among the issues facing these traditionally regulated firms is the nature and extent of future regulation at the state and federal levels.

The forces that have stimulated increased competition, however, differ substantially in each industry. In telecommunications, technological advances, especially microwave communications, have changed many of the characteristics that made the industry a natural monopoly. In natural gas, legislation adopted in 1978 to allow phased decontrol of wellhead prices has

A previous version of this chapter was presented at the Sixth Annual North America Meeting of the International Association of Energy Economists, November 6, 1984, and appeared in the Proceedings of that conference.

led to competition between natural gas and alternative fuels, as well as competition among alternative gas pipeline suppliers.

The long experience of the telecommunications industry in its transition from monopoly to competition may offer insights for the pipelines and their regulators into the problems, and possible solutions, associated with such a transition. Four issues are of primary concern to policymakers: network access and mandatory carriage; local bypass; preferential pricing; and affiliate relationships. Three other issues are especially relevant to corporate planning: partial deregulation; unbundling of services; and federal and state regulatory relationships.

This chapter discusses each issue in turn, with the objective of helping regulators, legislators, and the pipeline industry meet the challenges that lie ahead.

Network Access and Mandatory Carriage

When the Federal Communications Commission first permitted new interstate carriers to compete with AT&T in 1971 (after AT&T court appeals were resolved), it also required AT&T and its operating companies to interconnect with the new competitors. This regulation enabled these carriers to provide end-to-end service. Long battles in and out of court arose from AT&T's reluctance to provide adequate access. Throughout, AT&T complained that the new firms were free to serve only the most lucrative routes, thereby "cream-skimming" while AT&T had the obligations of a traditional public utility to provide nationwide service.

This experience bears some similarities to legislative moves being considered to require natural gas pipelines to sell transportation service to purchasers of gas when capacity is available. Indeed, some legislative proposals would extend the Federal Energy Regulatory Commission's (FERC) authority to require capacity expansion to provide additional transportation service. Mandatory carriage could allow customers to shift back and forth between pipeline system supplies and spot-market supplies, depending on relative prices. The troubling effects of such proposals are similar to the cream-skimming problems noted by AT&T: pipelines would be expected to continue honoring their public utility obligations to supply gas (on demand) as well as transportation, while competitors could move in and out at will without any such obligations.

One major difference does arise, however, between the reluctance of pipelines to support mandated access proposals and the reluctance of AT&T to interconnect. In the case of AT&T's reluctance to connect, anticompetitive conduct during the prolonged transition in telecommunications has been the primary issue. New carriers, such as Microwave Communications, Inc.

(MCI), have accused AT&T of anticompetitive practices related to access and have won court judgments against AT&T. For the pipelines, obligations of pipelines to purchase (or pay for) minimum quantities of gas from producers (take-or-pay oblications) are an important consideration. Depressed demand in the gas market has exposed many pipelines to high take-or-pay prepayments for gas that they could not take. This problem would be exacerbated if customers displaced pipeline sales with their own purchases and used pipelines only for transportation. In the long run, as current take-or-pay obligations are decreased, these difficulties will diminish. But with mandatory carriage, the cream-skimming problem could continue to nag both the industry and its regulators.

Local Bypass

Recent advances in telecommunications technology have made it possible for some businesses to bypass local telephone loops. Bypass is feasible with microwave links, cellular radio channels, cable, or other alternatives. Some users find bypass less costly than paying the share of fixed costs for the conventional local service allocated to them.

Bypass is also a potential problem for the end-use markets served by the interstate gas industry. Although practices vary from state to state, large gas users frequently bear a significant share of the fixed costs associated with local gas distribution and often must pay substantial utility (sales) taxes on purchases. Such practices encourage large industrial end users to bypass the local distributor and purchase gas directly from a pipeline or a producer.

Two distinctions between telecommunications and natural gas will affect the outcome of this issue in each industry. First, the advance of technology makes bypass prevention a problem for telecommunications, because technology advance can outpace bypass restrictions. It may be possible to prevent a particular type of bypass, but new possibilities are quickly developing. The prevention of local bypass is relatively straightforward for gas. Second, potential telecommunications bypassers generally do not contribute to improved load management. The potential bypassers for natural gas are off-peak customers who are important to efficient use of the transmission and distribution systems and who could leave the system entirely by switching to alternative fuels.

The distinctions between the industries suggest (1) that local bypass is easier to resolve for gas, but (2) that regulators must design gas rates that take into account the benefits that off-peak users can contribute to the system.

Preferential Pricing

Both industries have experienced debates over whether preferential pricing schemes designed to encourage customers to remain on the system are appropriate. AT&T's attempts to lower rates to certain users, in response to interstate competition, triggered concerns that AT&T was pricing these services below cost and compensating by hiking rates to its basic public telephone subscribers, who did not have alternatives. Rate hearings have dragged out over years, largely because of controversies about how fixed costs should be shared among multiple services. Most notably, AT&T's rates for Telpak (a discounted bulk private-line service aimed at large businesses) was declared unlawful by the FCC after fifteen years of hearings and investigations. Users, by then dependent on the bulk-rate service, protested and delayed withdrawal for another five years. These prolonged contests and the FCC's dissatisfaction with cost data from AT&T led elements of the FCC to prefer structural change as a way to control AT&T power.

Must the pipeline industry look forward to similar prolonged and costly proceedings as it seeks to adjust rates to fit a more competitive environment? Delays may not last fifteen years, because, among other reasons, the proportion of pipeline fixed cost to total cost (including purchased gas) is less in gas than in telecommunications, reducing the effects of alternative ways of allocating fixed costs. The debate over what share of fixed costs to allocate to customers who can switch off the pipeline system to other fuels will not be resolved easily. Regulators must assess when interruptible users are contributing enough to fixed costs to benefit all system customers. But the solution to this problem, which varies among pipelines depending on take-or-pay obligations and other factors, may be easier to determine than is the parallel problem in the telecommunications industry.

The political obstacles to an efficient solution could also remain significant. Most load-management factors for interstate gas favor large business customers—unlike telecommunications, in which residential users provide off-peak load. Moreover, litigation could delay the competitive moves of the interstate pipelines more than their less-regulated competitors. This advantage could pose a greater problem for gas pipelines than for AT&T, because gas markets are likely to be stable, whereas telecommunications markets are expanding rapidly.

Affiliate Relationships

In both industries, issues have arisen about producer-affiliate relationships. Western Electric, owned by AT&T, has been virtually the sole supplier of equipment to AT&T and (until the divestiture) to the Bell Operating Companies. Concerns about possible preferential treatment of Western Electric helped trigger an antitrust suit by the Department of Justice that led to a consent decree in 1956 and to a second suit that culminated in the recent divestiture of the Bell Operating Companies.

The situation faced by the pipelines stands in sharp contrast to that of Western Electric. First, interstate pipelines purchase only a small percentage of their gas from producer-affiliates (12 percent in 1983). Second, gas is a relatively homogeneous commodity, unlike the thousands of different types of telephone equipment, some tailored to the special needs of the Bell System. Therefore, comparisons between gas prices by affiliates and independent producers can be used to detect preferential treatment, if it exists at all, in contrast to the far greater difficulty of making comparisons across a vast and heterogeneous set of telecommunications products.

The smaller degree of vertical integration and the greater ability to identify comparable treatment of affiliates in natural gas than in telecommunications indicate that the telecommunications policy solution of divestiture would be unwarranted in natural gas.

Partial Deregulation

Regarded by the FCC as a dominant interstate carrier, ATT Communications (the recently renamed portion of AT&T, supplying long-distance services) remains subject to the full panoply of regulatory control while competitors benefit from streamlined procedures. Thus, any new tariffs ATT Communications files must be supported by detailed evidence of their justness and reasonableness, evaluated in lengthy and formal proceedings. Other firms, judged to be nondominant, can more easily file tariffs in accordance with changing market conditions. Therefore, ATT Communications not only faces cream-skimming, it is also handicapped by asymmetrical regulatory oversight but is expected to hold to public utility obligations. Recognizing this problem, the FCC is investigating the possibility of reducing the regulatory burden on ATT Communications, perhaps on a phased basis.

Under several recent proposals, pipelines could encounter a similar world of regulatory disparities. Some proposed legislation in the last session of Congress would have allowed intrastate pipelines to engage in interstate purchases and transportation of gas without being subject to the same federal controls as are interstate pipelines. For instance, interstate pipelines would have serious disadvantages in competing for gas supplies if the requirements for prudence review or mandatory filing of contracts included in some legislative proposals were adopted. Intrastate pipelines, brokers, and industrial users seeking gas supplies would not be subject to the same review. Moreover, purchased gas adjustment (PGA) filings already provide competitors with useful information about the supply costs and sources of interstate pipelines. Perhaps the greatest disadvantage comes when pipelines must compete today with unregulated fuel oil suppliers. Pipelines could find themselves too burdened by regulators to compete effectively and too exposed to competition to retain franchise protection for markets. Unfortunately, in contrast to the FCC's activities, the FERC is not considering reducing the regulatory burdens on interstate pipelines.

Unbundling of Services

Until the late 1970s, telephone services generally were provided on a bundled, full-service basis. The monthly charge for residential telephone service, for example, included provision and full maintenance of the telephone (owned by the local telephone company), along with local calls. But the growth of competition in recent years is forcing each component of service to be offered separately. Consumers are now free to buy their own telephones, marketed by a host of alternative suppliers, while purchasing only transmission services from telephone companies.

The lesson here for the pipelines is that they, too, may have to unbundle in response to competition. Until recently, most gas transactions involved full-service purchases, which included not only transportation but brokering, storage, load balancing, and sometimes gas production and marketing as well. Some pipelines have already decided that this practice is no longer sustainable in the face of increased competition. They are offering, instead, a variety of separate services from which customers may choose. Although some firms may elect to maintain traditional full service, others now offer a range of services tailored to specific market segments.

Federal and State Regulatory Relationships

In telecommunications, the federal and state jurisdictions are split by the type of service offered. Interstate long-distance rates are regulated by the Federal Communications Commission. Local and intrastate long-distance rates are regulated by the states.

Therefore, interstate telecommunications carriers face less uncertainty about how their actions will affect end-user prices and, hence, the marketability of their product, because the relationship between their actions and end-user prices is affected by only one rather than two layers of regulatory jurisdiction.

Interstate gas companies need to take into account the key effect that both federal and state regulatory decisions have on natural gas marketability.

Pipelines also need to pay greater attention to the regulatory process at the state level. For interstate gas, state public utility commissions set final prices for most end users, while federal regulators set tariffs governing the city-gate price. In essence, the states add on their own local costs to the federally regulated price. Pipelines are dependent on state as well as federal regulation to allocate costs in a manner that enhances gas marketability.

Conclusions

A common theme in this chapter is the belief of telecommunications and natural gas policymakers that the ability of the traditional regulated firms to control access to the transmission system is a key source of market problems. Proposed solutions in both industries currently require that traditional suppliers remain under public utility obligations to serve existing customers while potential competitors are allowed to pick and choose markets.

When AT&T faced this market situation, it decided to settle with the Department of Justice and triggered the latest phase of telecommunications competition. Now pipelines must ask whether the broad framework for regulation of natural gas pipelines assures that pipelines can respond quickly and flexibly to competitive market forces.

In telecommunications, the treatment of traditional suppliers, though burdensome, may be more manageable because the industry is benefiting from rapid technological advances, combined with continued strong growth in market demand. Under these conditions, a traditionally regulated firm may be able to survive as a carrier of last resort, despite loss of customers to other suppliers.

In contrast, the interstate natural gas pipeline industry is blessed neither with rapid technological advances nor with continued strong growth in market demand. Were it subject to the restructuring being pursued in the telecommunications field, it would have far greater difficulty in discharging its public utility obligations, causing possible reductions in service availability to some customers and escalating gas prices for those unable to switch to alternative sources.

Other key points also emerged from this analysis. First, rate design reform has not worked for telecommunications because of the complexity of affiliated transactions and the high proportion of fixed and joint costs, although rate design reform offers great potential for resolving gas market problems such as local bypass and the potential loss of large industrial customers. Second, state PUCs have a crucial role in establishing gas marketability, because they set end-user rates. Pipelines therefore need to become participants in PUC decision processes. Finally, unbundling provided full-service telecommunications firms with a response to competition, and unbundling is creating new business opportunities for some pipelines.

21

Assessing the Feasibility of Modeling the Economic Impacts Associated with Changing Carrier Access and Customer Line Charges: A Generic Study of the Southern Region

R. Carter Hill
Louisiana State University
Albert L. Danielsen
David R. Kamerschen
University of Georgia

The Federal Communications Commission (FCC) Telephone Access Charge Plan (FCC Docket No. 78-72) and subsequent rulings provide for the implementation of a customer line charge to defray the non–traffic-sensitive costs of the local telephone loop and central office. Higher customer line charges would permit a reduction in common carrier access charges, which are derived from toll traffic, and a reduction in toll rates. In essence, the FCC access charge plan would be a step toward a more cost-based and sustainable system for rendering telecommunications services.

Given the changes advocated by the FCC, the question of the economic effects of the impact of the plan naturally springs to mind. More specifically, what are the probable economic effects of specified changes in customer line and common carrier access charges? This question was the primary focus of a study conducted by Wharton Econometric Forecasting Associates [hereafter referred to as Wharton] that was released in November 1983 (see Wharton, 1983). The Wharton study, based on a macroeconomic model of the United States, provided estimates of the effects the FCC plan would have on whole economy.

The Wharton study indicated that lower common carrier access charges may be expected to lower toll rates, increase the overall quantity of toll services demanded, promote the substitution of long-distance telecommunication services for more costly travel and related business expenditures, and increase real personal incomes. In essence, a more nearly cost-based system of telecommunications pricing will tend to promote a more efficient utilization of the nation's resources.

Although the exactness of the Wharton study has been called into question (primarily because the magnitude of the anticipated economic impacts is very large and because the impacts are exceedingly widespread), there can be little doubt that the main thrust of the study is correct. The results of an economic impact assessment based on a formal econometric model obviously depend on the validity of the structural equations and the parameter values. However, the Wharton exercise was valuable because it focused attention on the fact that there are benefits as well as costs associated with changing local and toll rates.

Our study of the nine southern states is based on the premise that it would be educational for researchers, regulatory commissions, consumers, and telephone companies to undertake studies of the Wharton type for one or more of the states. More specifically, the primary purpose of the study is to assess the feasibility of undertaking studies such as the one conducted by Wharton for one or more of the nine southern states we selected. The specific questions that will be addressed are

1. How many of the targeted southern states have macroeconomic forecasting models that could be appropriately modified to allow an assessment of the economic effects of changes in carrier access and customer line charges?
2. What is the general nature and structure of each state model, and, assuming that an appropriate procedure is developed to utilize a specific model to make economic impact assessments, what information could each model provide about the effects of changes in carrier access and customer line charges?
3. If it were deemed advisable to modify the state models, what modifications would be the more feasible and less costly (that is, cost-effective)?
4. Again, assuming it is deemed advisable to modify the state-specific models, what would be the alternative ways to accomplish those modifications, and what approach or approaches should be adopted?

This chapter is organized around these questions. The next section contains a brief discussion of the Wharton methodology, along with an assessment of its potential applicability to the state level. We then provide a more comprehensive explanation of the product and income approaches to macroeconomic modeling and a general discussion of the procedures that would be required to modify each type of model to evaluate the effects of changing telecommunications prices. The next section is devoted to specifying a general telecommunications subroutine (or model) with a residential and business component. The knotty issues involved in estimating the effects of changing residential and business telecommunications prices and usage are also discussed in that section. We then provide a list of the states included in the study and a report on the status of their macroeconomic forecasting efforts. The overall structure of each state model is specified, along with an assessment of what information could be derived from them. The chapter concludes with an assessment of the technical and strategic issues that will have to be addressed if and when the models are modified to assess the economic impact of changes in carrier access and customer line charges.

The Wharton Study

The Wharton approach is described in two reports. The basic methodology is described in a Wharton report released in January 1984. This was the methodology used to produce the basic Wharton (1983) study of the effects of the FCC Access Charge Plan on the U.S. economy. The procedure used by the Wharton researchers is straightforward. They first developed a base case and an alternative scenario that assumed that the FCC acess charge plan was adopted. Alternative price paths for local service and four categories of long-distance service—message toll service (MTS), inward-WATS (800 service), outward-WATS, and private line service—were assumed for each scenario for each of the years 1984 through 1988. Separate price paths were developed for business and residential customers. Average values of the own-price elasticities of demand for local and toll calls, derived in an independent study by Lester D. Taylor (1980),[1] were used to calculate expected changes in the quantities of telecommunications services that would be demanded by residential and business customers under each scenario. These values were then fed into the Wharton long-term model to derive the changes that are expected to occur in the industries that provide inputs or use the outputs of the telecommunications industry.[2]

The major results of the Wharton study may be summarized as follows:

1. The increase in the fixed portion of the telecommunications service cost will lead to a slight reduction in the use of local service by both residential and business customers.
2. The reduction in long-distance rates will lead to a large increase in long-distance use by both residential and business customers.
3. Overall, it is predicted that there will be a decrease in the average (local and toll combined) price for telecommunications services.
4. As a result of these changes, they then predict lower consumer prices (that is, a lower CPI), lower interest rates, higher disposable income, an

increase in net investment, an increase in net exports, an increase in gross national product (GNP), and lower unemployment.

The Wharton method is powerful since it captures interindustry linkages and allows one to trace the impact of alternative policies directly to specific input and output categories. However, this very strength is, of course, why the Wharton procedure is not particularly appropriate for use at the state level. State macroeconomic models typically do not have input-output structures that are as sophisticated as those of the Wharton model. Therefore, interindustry effects and impacts on the various demand categories cannot be captured in this way. In the next section of this chapter, we examine the structures of existing state models and review ways that they might be used to answer the questions that were posed at the beginning of the chapter.

Alternative Approaches to Macroeconomic Modeling

Background of Modeling Efforts

A general description of the econometric approach is necessary to an understanding of modern macroeconomic models and how they can be used to analyze the state-level impacts of decisions such as those contained in FCC Docket No. 78-72. Econometrics is a subdiscipline of economics that describes the ways in which statistical procedures and modeling can be combined to quantify economic relationships. However, econometrics is more than the application of statistical techniques to economic data, since it implies that there is an underlying economic structure that can be modeled, quantified, and tested.

There are two major outputs of econometric analysis. First, a quantified model is produced that can be used to forecast. Second, in the process of specifying the model, the theoretical foundations of the model are statistically tested. The process of submitting theoretical models to empirical tests using econometric techniques is basic to the way most modern economic research is carried out. State and regional macroeconomic models are excellent examples of how econometrics is presently used to forecast and test. These models are a natural vehicle for analyzing the state-specific impacts of a change in carrier access and customer line charges.

State and regional econometric models have been constructed with the dual functions of linking the regional economic outlook to that of the national economy and of forecasting the effects of changes in state and national policies. These models are often built as satellites to a model of the U.S. economy and assume that causation is one-way (that is, from the U.S. economy to the state economy), with no feedback effects. The basic structures of the more parochial macroeconomic models fall into two categories: those based on gross regional output and those based on gross regional income. In this section of the chapter, we present a general description of these two types of models and discuss how each may be used to evaluate the effects of changing telecommunications prices.

The Product Approach

Gross state product is the sum of gross output by each sector of the state's economy. Models based on gross state product typically consist of (1) an output block, which models output in the various sectors; (2) an employment block, which estimates employment, wage rates, and personal income by sector; and (3) a tax block, which estimates tax revenues, given the tax bases and rates. The flowchart in figure 21–1 provides an illustration of a gross state (Mississippi) product model (see Adams, Brooking, and Glickman, 1975).

Changes in telecommunications prices affect the state economy in several ways. The most direct effects are the result of changes in the overall U.S. economy, which are exogenous to the state models.[3] This can be handled by inputting the results of the Wharton study directly into the state model.

In addition to these exogenous effects on the state economy, there are also impacts due to endogenous changes in telecommunications prices within the state:[4]

1. The effects on the output of the telecommunications industry follow directly from one of the most basic of all economic principles: the law of downward-sloping demand. The lower overall price of telecommunications services will lead to an overall increase in usage. To estimate these effects, the elasticities of demand for both residential and business customers for local and toll calls must be ascertained. Detailed discussion of this topic is provided later in this chapter.
2. Another change will be carried through the various industrial sectors in the economy. The lower telecommunications prices will reduce the cost of production and lower the prices of output, thus increasing the real value of the output. More will be said about these linkages later.
3. The state-level increases in real output in the telecommunications industry and in the other business sectors add to the exogenous effects of the U.S. model to drive the state model. The growth path of the variables that are endogenous to the system (that is, the endogenous variables) can be evaluated by substituting alternative values of the variables that are exogenous to the system (the exogenous variables). These, of course, include the level and structure of telecommunications prices.

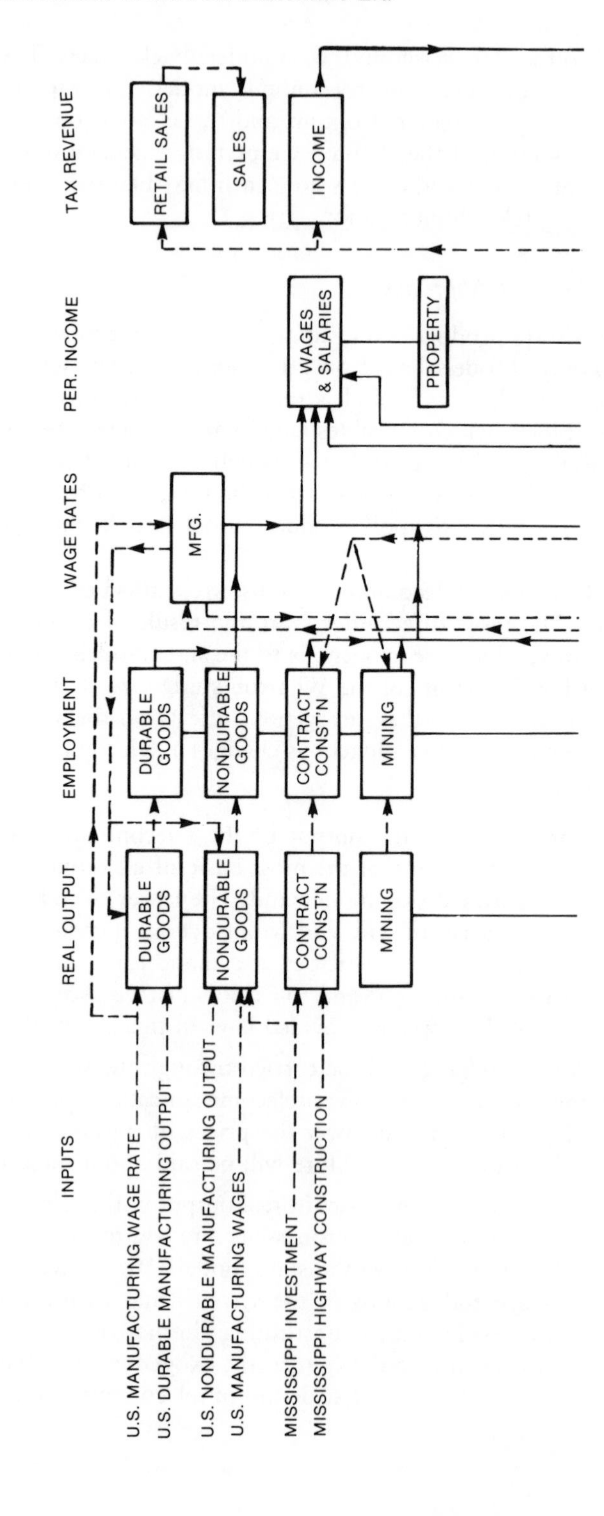

TAX REVENUE
RETAIL SALES
SALES
INCOME
PER. INCOME
WAGES & SALARIES
PROPERTY
WAGE RATES
MFG.
EMPLOYMENT
DURABLE GOODS
NONDURABLE GOODS
CONTRACT CONST'N
MINING
REAL OUTPUT
DURABLE GOODS
NONDURABLE GOODS
CONTRACT CONST'N
MINING
INPUTS
U.S. MANUFACTURING WAGE RATE
U.S. DURABLE MANUFACTURING OUTPUT
U.S. NONDURABLE MANUFACTURING OUTPUT
U.S. MANUFACTURING WAGES
MISSISSIPPI INVESTMENT
MISSISSIPPI HIGHWAY CONSTRUCTION

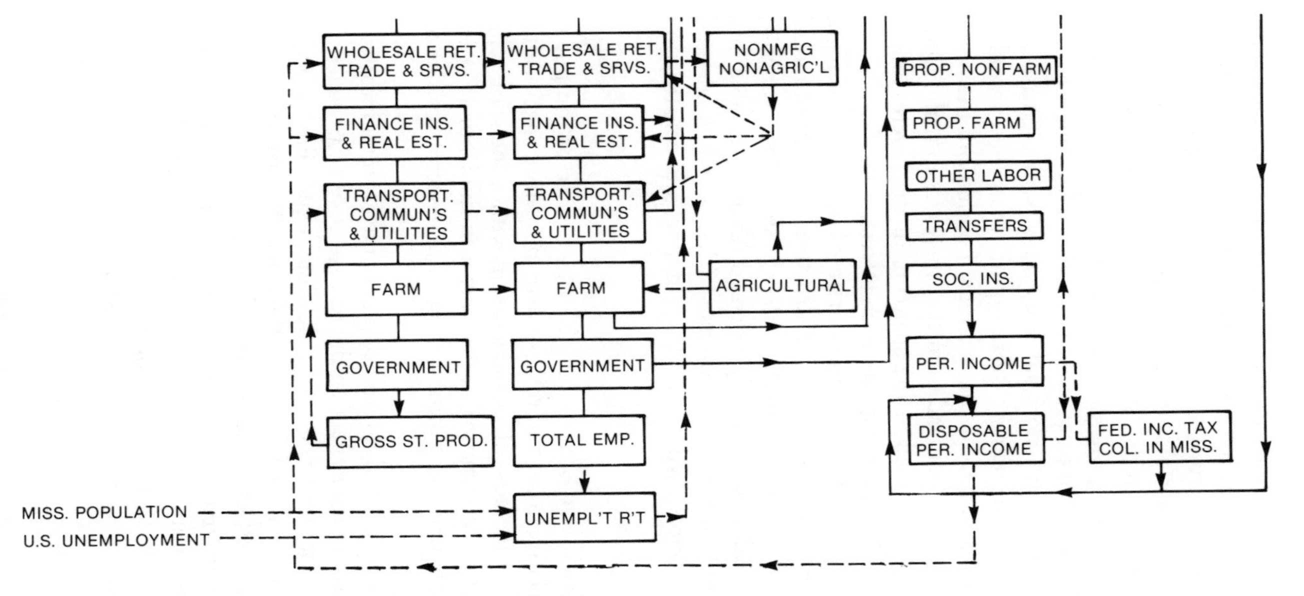

Figure 21-1. Structure of the Mississippi Model

Source: Gerald F. Adams, Carl G. Brooking, and Norman J. Glickman, "On the Specification and Simulation of a Regional Econometric Model: A Model of Mississippi," *Review of Economics and Statistics* 57(1975), Figure 1, p. 288.

For example, lower telecommunications prices, which are exogenous to the system, will cause predictable changes in a number of the variables that are endogenous to the system. This is especially true with regard to the direction of expected changes, but also, to lesser extent, to their magnitude. Because of lower overall telecommunications prices, it is reasonable to expect that the cost of living will grow more slowly and that nominal wage rates, especially those that are linked to a cost-of-living adjustment clause, will also grow more slowly. Thus, nominal wage rates will increase less rapidly and real wage rates will increase more rapidly than they would in the absence of price changes in telecommunications. In addition, there will be secondary and tertiary effects that lead to further increases in real output. These linkages are already embodied in the structure of the macromodels, and no modifications are required to capture such effects.

To indicate more precisely how an output-oriented state model might be used to capture macroeconomic impacts, the structure of a typical model can be examined more closely. It is convenient to classify, albeit roughly, the output sectors of a model as externally oriented (that is, durable and nondurable manufacturing) and internally oriented (that is, wholesale and retail trade, finance, and services), with agriculture, mining, and construction treated separately.

Externally Oriented Sectors. For each externally oriented sector, output is assumed to depend on relative costs and relative demand. More specifically, sectoral output depends on the cost of manufacturing within the state relative to the average cost of manufacturing in that sector in the United States. It also depends on the state demand for the output of that sector relative to the total U.S. demand.

With respect to the issue at hand, overall changes in telecommunications prices are not likely to have much impact on the relative costs of production within a particular state, because telecommunications expenditures are a small portion of total manufacturing costs and the price changes are occurring nationwide. The major effects of telecommunications price changes on manufacturing may be expected to come from changes in U.S. economic activity (demand for sector output) and from reductions in overall costs of production due to the lower average price of telecommunication services. Therefore, the specific questions that need to be asked about the externally oriented industries are

1. What changes in sector output will occur because of the increase in U.S. demand?
2. How much to the costs of production fall in each industry?

Internally Oriented Sectors. The effects on the internally oriented industries are assumed to be related to the costs of producing the output of the sector and the demand for that output as related to the level of state income. On the cost side, telecommunications expenditures are a relatively more significant portion of the sectoral costs, as contrasted with an externally oriented sector; thus, one may expect short-term increases in the real value of output. There may also be long-term substitutions of telecommunications inputs for other inputs, especially for labor. Thus, the relevant questions for these industries are

3. How much will real income in the state change?
4. How much do the costs of production change in each industry?
5. What long-run input substitutions will occur?

Application to the State Models. The answers to questions 1 and 3 are already built into the state models. Changes in sectoral output for the U.S. economy are exogenous to the state models, and these effects have been specified. State income is endogenous to the models, and feedback effects on the internally oriented sectors are taken into account by the structure and equations in the models. The answers to questions 2, 4, and 5, however, require a careful study of the inputs into each sector and their substitutability.

The issues of input substitutability are especially important in these models as they relate to the demand for labor. In an output-oriented model, the results from the output sector are used as inputs into the labor market block. Specifically, the demand for labor in each sector depends on the real output in that sector and the real wage rate. However, as telecommunications prices fall, we expect a substitution of telecommunications for labor services. The question for each sector is how extensive these substitutions will be and how long they will take. One way to approach this problem, which will also be applicable to models based on gross regional income, is to expand the factors that affect sectoral demand for labor to include the price of telecommunications services—which can serve as either a net substitute or complement to other inputs, depending on the industry.

The Income Approach

An alternative to using models based on gross regional product is to use those based on gross regional income. The advantages of this approach include the availability of quarterly data on income and the ability to provide continuing and timely forecasts for policymakers. The starting point for a model based on regional income is a block of equations representing the labor demand and labor supply in each sector of the economy. The labor market block is used to provide forecasts of employment and wages. These forecasts, in turn, are used to forecast personal income and state taxes. Once again, these blocks of the state model are satellites of a U.S. macroeconomic model. National exogenous variables—such as GNP, price indexes, interest rates, unemployment rates, and industrial production—are assumed to influence the state economy directly. As a prototype, the structure of the Delaware model is given in figure 21–2 (see Latham, Lewis, and Landon, 1979).

The driving force in an income-oriented model is the labor market block, and it is in this block that the effects of telecommunications price changes appear. To discuss these effects, it is once again convenient to break the state economy into sectors that are externally and internally oriented.

Externally Oriented Sectors. These sectors respond to changes in the level of national economic activity through changes in the national exogenous variables, such as increases in GNP or industrial production, that serve as inputs to the state models. In particular, changes in the national exogenous variables will change the demand for the products of externally oriented sectors, and these changes appear in the model as changes in the demand for labor.

Internally Oriented Sectors. For sectors that are internally oriented, the demand for labor depends on state income, which is determined by aggregating total sector wage bills plus other income. In turn, the total sector wage bills are products of average hours of employment in each sector, multiplied by the sector wage rate. The average hours of employment variable in each sector depends on the level of economic activity within the state, as measured by state income. Thus, state income and employment are determined jointly.

In addition, the changes in the national and state telecommunications prices affect both internally and externally oriented sectors in two other ways. First, as the overall costs of production fall, prices of consumer goods fall and real income rises. This rise in real income affects the demand for labor in both internally and externally oriented sectors as state and national incomes increase.

The lower telecommunications prices also have an intrastate impact as producers adjust their (long-run) input mix to minimize costs. These changes show up in the demand-for-labor equations. Changes in the prices of substitute and complementary inputs do change the optimal (or best) input mix, which, in turn, shifts the demand for labor.

The best way to incorporate these effects is to reestimate the various demand-for-labor functions with the prices of competing inputs, such as telecommunications services, included. Alternatively, if the elasticities of labor demand with respect to telecommunications prices were available from some other source, the intercepts of the existing labor demand functions could be adjusted and reestimation could be avoided.

Figure 21-2. Structure of the Delaware Model

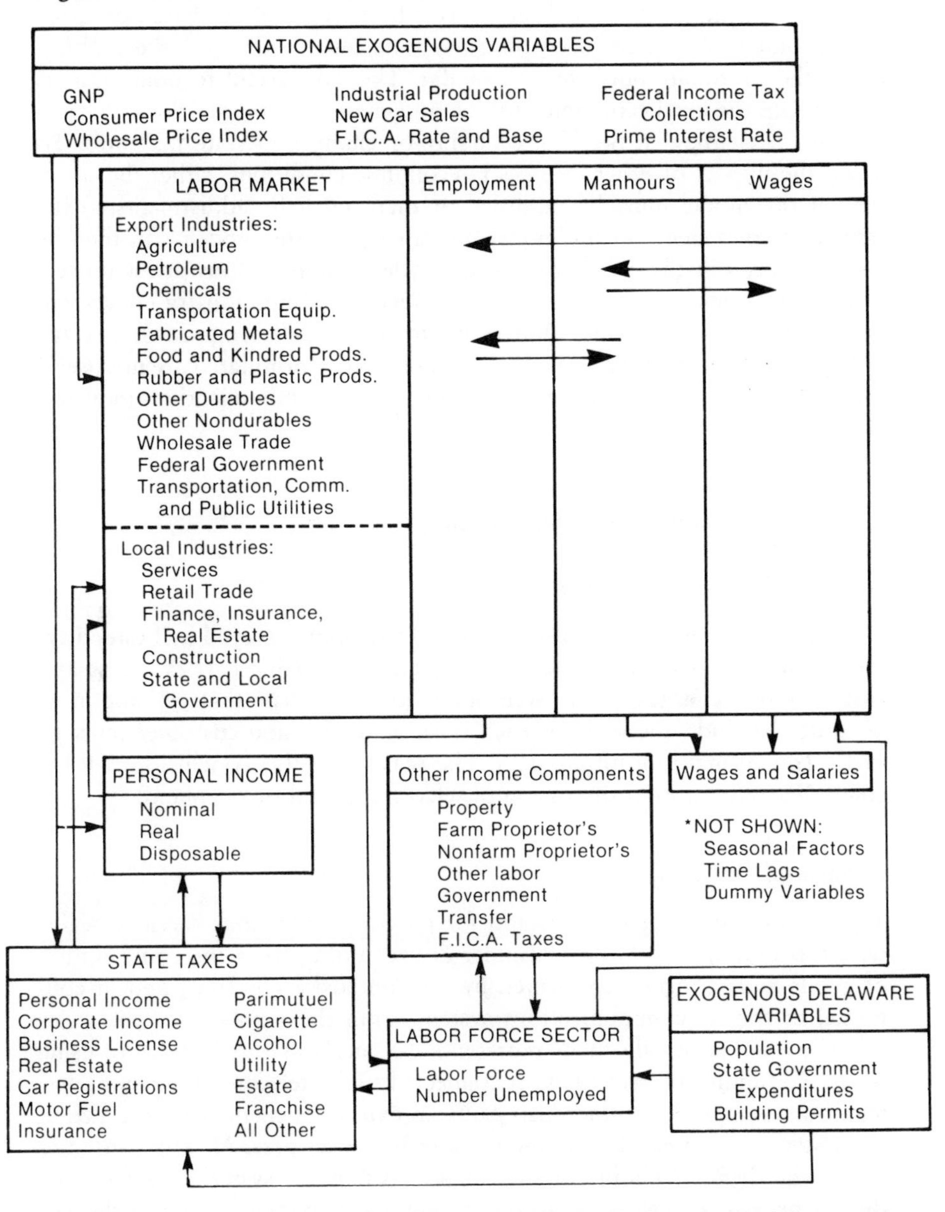

Source: William R. Latham, Kenneth A. Lewis, and John H. Landon, "Regional Econometric Models: Specification of a Quarterly Alternative for Small Regions," *Journal of Regional Science* 19(1979), p. 5.

Summary

Currently existing state models are of two basic types: those based on gross state product and those based on gross state income. Despite these differences, their ultimate purposes are similar. They are useful to policymakers in assessing the local economic outlook and changes in that outlook that are a result of changes in the U.S. economy or changes in state tax rates. To assess the impact of other changes in economic conditions, either the structure of the model must be modified or there must be adjustments to the estimated coefficients in the model. In either case, the models can then be used to trace out the effects of changing telecommunications prices on sectoral employment, income, and state tax revenues. Interindustry effects will be captured so long as the adjustments to the model recognize that sector differences exist. However, the models as presently formulated are not useful for the purpose of examining effects within the telecommunications industry. We now turn to this task.

The Telecommunications Subroutine

General Theoretical Formulation

The theory of telecommunications demand has been surveyed by Lester Taylor (1980). To assess accurately the impact of changing carrier access and customer line charges at the state level, it seems clear that estimation of separate demand functions for each class of service and customer must be carried out using appropriate data. The reason for this conclusion will become clear after a brief discussion of telecommunications demand.

Residential Demand

The analysis of residential demand for telecommunications services begins with the assumption that a consumer derives utility from the consumption of goods and services, represented by the composite good (x), and, if connected to the telecommunications system, from the number of calls made (q). The number of calls made represents the purely private benefits of using the telecommunications system. If connected to the telecommunications system, however, the consumer also gains an external benefit from the size of the system, as represented by the number of subscribers, N. The consumer makes the decision about whether or not to join the system by evaluating the net benefits of belonging to the system and making calls, given the line charges required to access the system (r), the price of a call (v), the price of the composite good (p), and the consumer's income (y).

Using standard economic analysis, it can then be shown that the demand for the number of calls and the composite good depend on the price of a call, the price of the composite good, the number of subscribers to the telecommunications system, and the amount of income minus the customer line charge. We express this algebraically as follows:

$$q = q(v, p, N, y - r) \tag{21.1}$$

$$x = x(v, p, N, y - r) \tag{21.2}$$

If this analysis is extended to the entire population, the population demand for telephone calls (Q) will depend on the same factors as before, but with population income (Y) replacing individual income. Thus:

$$Q = Q(v, p, r, N, Y) \tag{21.1}$$

The only difference between the equations for market and individual demand is that the customer line charges are included separately in the market demand equation rather than as a subtraction from income.

The demand for access to the system by the population will be manifested by the proportion of the population that is connected to the telecommunications system. This proportion is $m = N/M$, where, as before N represents the number of individuals subscribing to the system and M is the size of the population. Each individual is assumed to evaluate the net benefits of belonging to the system, and N is simply the number who décide that the net benefits are positive. The factors that determine this proportion are the price of telephone calls, the price of other goods, the customer line charges, the number of subscribers and aggregate income:

$$m = N/M = m(v, p, r, N, Y) \tag{21.4}$$

Equations 21.3 and and 21.4 represent, respectively, the basic demand by residential consumers for telephone calls and for access to the system. We will now elaborate on these basic relationships.

Option Demand

Another factor that affects the demand for customer access to the system is an option demand. The decision to subscribe provides the consumer with the option of making many calls, regardless of whether or not those calls are actually made. The benefits of the calls that might have been made cannot be quantified directly, but the central idea can be incorporated into the demand function in several ways. For example, the potential calls might be made for several purposes, one of the more important of which would be emergency calls. Generally, this option may be more important in rural area because of the lower population density. If this is true, an additional variable in the individual demand for access might be whether the individual is a resident of a rural or an urban area. Consequently, a variable measuring the proportion of the population residing in a rural area might be added to equation 21.4.

Multipart Pricing and the Duration of Calls. The next characteristic of telecommunications demand to be introduced are the multipart pricing and duration aspects of a phone call. Multipart pricing can be taken into account by breaking the price of a call into the customer line charges required to access the system (r), the charge for the initial period of the phone call (v_0), and the charge for each overtime period (v_1). The measure of telecommunications services is now the duration of the call, represented by q_i, instead of the number of calls. Aggregating over all consumers in the population, or summing the q_i, we obtain

$$Q = Q(v_0, v_1, p, r, N, Y) \tag{21.5}$$

$$m = N/M = m(v_0, v_1, p, r, N, Y) \tag{21.6}$$

In these models, changes in v_1 measure the traditional price effect, assuming that calls last into the overtime period, and changes in v_0 will cause an income effect. The effects of both of these changes are negative.

An alternative approach to investigating the effect of prices on duration is to break Q in equation 21.5 into two components: the number of calls (Z) and the average duration of a call (t). That is,

$$Q = Z = t \tag{21.7}$$

The study of Z and t can then be made, using the same framework as the demand for customer access and the demand for use. The decision to make a call corresponds to the demand for customer access, and the duration of the call corresponds to use. Then v_0 represents the buy-in price of a call and v_1 represents the running cost of use. Thus, v_1 will be most relevant for explaining t, but both v_0 and v_1 will affect Z, since both factors affect the net benefit of making a call. The demand equations for Z and t can therefore be represented as

$$Z = Z(v_0, v_1, p, r, N, Y) \tag{21.8}$$

$$t = t(v_0, p, N, Y) \tag{21.9}$$

Note that r has been omitted from equation 21.9. The reason is that although the duration of a call clearly depends on the customer paying the line charges required to access the system, r, changes in the line charges will have a negligible effect on income.

The Length of Haul. The next complication to be introduced is that the price of a call depends on the distance the call traverses (that is, the length of haul). Let us consider the one-way toll volume, measured in call minutes between two sites, or systems. Let $Q(i, j)$ represent the number of call minutes from system i *to system* j. Correspondingly, let $Z(i, j)$ be the number of calls from i to j and $t(i, j)$ be the average duration of a call from i to j. Then

$$Q(i, j) = Z(i, j) \times t(i, j) \tag{21.10}$$

The factors affecting the number of calls, $Z(i, j)$, are

$$Z(i, j) = Z\{v_0(i, j), v_1(i, j), p, r, (i), N(i), N(j), Y(i), d(i, j)\} \tag{21.11}$$

where $v_0(i, j)$ = price of initial period of call from i to j
$v_1(i, j)$ = price of overtime period from i to j
p = price of other goods and services
$r(i)$ = customer line charges for access to the ith system
$N(i), N(j)$ = numbers of subscribers in systems i and j, respectively
$Y(i)$ = population increase in system i
$d(i, j)$ = distance between system i and j.

The effects of system sizes are expected to be positive, since the greater the number of subscribers in each system, the more likely it will be that messages will be transmitted between them. The effect of distance is expected to be negative, since the farther apart the sites are, the lower the probability that individuals will be acquainted.

The mean duration of calls from system i to system j, $t(i, j)$, is explained by

$$t(i, j) = t\{v_1(i,j), p, N(j), Y(i), d(i, j)\}, \tag{21.12}$$

with the variables defined as for equation 21.11.

Time-of-Day Pricing. A final variation is to incorporate time-of-day pricing. This is easily done by estimating separate price effects for, say, day and night calls from system i to system j by entering into equations 21.11 and 21.12 the relevant day and night rates (or rate differences) for the initial and overtime periods.

Availability and Usefulness of Data. The demand for local telecommunications service and the demand for customer access to the system are difficult to separate, since most local service is not measured. The demand for customer access is ideally studied by using household data with complete household demographics. Although it has not been previously stressed, not only can the effects of the various telecommunications prices on subscription rates be studied, but questions regarding household size, age composition, income, residential location, and so forth, can also be investigated with data of such detail.

There are a variety of statistical methods for the evaluation of dichotomous, or yes-no decisions. Each of them allows estimation of the effect on the probability of subscription by a household with specified characteristics, given a change in a particular explanatory variable, such as the customer line or access charges, r. Alternatively, if only aggregate data are available, customer line charge equations such as equation 21.4 can be estimated, but the detailed effects of demographic changes will be obscured.

To study the demand for toll calls, either intra-LATA or inter-LATA, a pair of equations such as equations 21.11 and 21.12 are appropriate. These specifications are expressed in terms of aggregate variables, most appropriately at the city level, although the use of more aggregated data is conceivable. The use of aggregate data for these equations has some advantages over the use of individual household data. For individual households, the presence of multipart pricing means that multiple equilibria are possible, so the demand functions for telecommunications service may not have an analytic form. The use of aggregate data obviates that problem.

Business Demand

The estimation of business demand is much more difficult conceptually than the estimation of residential demand. The reason is that the business demand for telecommunications services is a derived demand that is based on profit-maximizing behavior. To begin reasonably, the problem must be disaggregated, at least to the sector level. Within each sector, data on usage, basic inputs, and value of output for samples of firms must be obtained. Given that such data are available, sectoral demand for customer access and intra-LATA and inter-LATA service can be modeled using the same basic reasoning as outlined for the residential customer, with appropriate modifications of population and income variables to reflect the interdependencies within the economic sectors.

For example, the analogue to equation 21.11 for, say, the financial sector would depend on the number of similar or related firms in systems i and j and an income or earnings variable, not only for system i but also for system j, to reflect the level of sector activity in system j. Other attraction or gravity-type variables, such as mail volume between systems *i* and *j*, should be considered as potential predictors. At this point, very little else can be said about the modeling of business demand, because such efforts are in their infancy, and the literature provides relatively little guidance on the issues.

Implementation of the Telecommunications Subroutine

In this section, we specify the concrete steps required to implement the telecommunications subroutine that we have expressed in theoretical terms. The estimation of the residential consumer demand equations for telecommunications services at various levels of aggregation will first be considered. Then the same will be done for business demand. Finally, the problems and prospects for integrating these results with state macroeconomic models will be discussed.

Residential Demand Equations

The first part of the model to be considered is residential consumer demand. Residential demand should be broken into a decision to subscribe to the system and a decision to make calls. To make specific the nature of the tasks involved, consider the theoretical structure given in equations 21.5 and 21.6.

The decision by a given household to join the telephone system is a yes-no decision. Therefore, for each household, the question is simply whether or not the household belongs to the system. From the point of view of the whole system, then, the variable that needs to be explained is the proportion of the population that subscribes to the system. In a modeling context, this proportion reflects the probability that a household will belong to the system.

Statistical models designed to estimate the effects of explanatory variables on the probability of system membership are of two common forms, called *logit* and *probit* models. (For a description of these models, see, for example, Judge, et al. 1982, chapter 18.) The basic idea of these models is as follows: A sample proportion m is observed, which we assume is related to a true population proportion M by the relation $m = M + e$, where e is a random disturbance. The true population proportion of subscribers is taken to be a function of explanatory variables such as those in equation 21.6, and depending on whether logit or probit analysis has been chosen, the functional relationship is a nonlinear one, based on the logistic or normal cumulative distribution function (F), respectively, as follows:

$$M = F(\beta_0 + \beta_1 v_0 + \beta_2 v_1 + \beta_3 p + \beta_4 r + \beta_5 N + \beta_6 Y) \quad (21.13)$$

This model can be estimated by generalized least squares, given data on the sample proportions of subscribers in various cities and measures on the corresponding explanatory variables. The sign of an estimated parameter indicates the direction of the effect of the corresponding explanatory variable on the proportion of the population belonging to the system. We expect the signs of β_1 through β_4 to be negative and those of β_5 and β_6 to be positive. The estimated coefficients can also be used to predict the magnitude of the change in the population proportion M, given a unit change in an explanatory variable. This change is

$$f(\beta_0 + \beta_1 v_0 + \beta_2 v_1 + \beta_3 p + \beta_4 r + \beta_5 N + \beta_6 Y) \quad \beta_{-j}) \qquad (21.14)$$

where $f(x)$ is the probability density function associate with F and β_j is the coefficient associated with the variable whose magnitude has been changed. To evaluate this expression empirically, the values of the estimated coefficients must be inserted as well as representative values of the explanatory variables. These representative values might well be set to be average values for different regions of the state or different service areas, so that the changes in specific subscribership rates can be predicted.

The second decision to be discussed, which is conditional on the decision to subscribe to the system, is the decision about system usage. In effect, we are concerned with estimating a demand equation for telephone use. Unlike other common situations, the companion supply equation will not be specified and estimated, since, for practical purposes, the supply of services is infinitely elastic at prevailing prices. That is, given the price of telephone usage, the customer can expect a telephone supplier to accommodate any and all demands for service.

As noted in the preceding section, the residential demand equation can be estimated in various levels of detail. For purposes of discussion, consider equation 21.5. The dependent variable is the number of minutes of telephone services used by a population—say, a city, as a function of the explanatory variables. As is common in many demand studies, we assume the relationship to be linear in the logarithms, and given by

$$\begin{aligned} \ln Q = \Phi_0 &+ \Phi_1 \ln(v_0) + \Phi_2 \ln(v_1) + \Phi_3 \ln(p) \\ &+ \Phi_4 \ln(r) + \Phi_5 \ln(N) + \Phi_6 \ln(Y) + e \end{aligned} \qquad (21.15)$$

where $\ln(x)$ is the natural logarithm of x, the Φ_j's are unknown demand parameters to be estimated and, given the logarithmic specification, are actually interpreted as demand elasticities, and e is a random disturbance. For this equation to be estimated at the city level, data on the variables in the equation must be available over time and perhaps for several cities. Economic theory suggests that the expected signs of Φ_1 through Φ_4 will be negative, whereas the signs of Φ_5 and Φ_6 will be positive. The values of the estimated coefficients, which can be obtained by ordinary least squares if the basic regression assumptions hold, indicate the percentage change in the value of *Q*, minutes of telephone services used, given a 1 percent change in the corresponding explanatory variable.

One of the basic assumptions of the specification in equation 21.3 is that the coefficients are the same for all observations in the sample. To increase the chances that the assumption will hold, equation 21.3 should be estimated separately for intra-LATA, inter-LATA, and interstate calls. One would expect that the elasticities of demand would increase through these classes of service, partially because of the distances involved. To take into account the length of haul, the specification in equation 21.10 should be used.

The local calls category has been omitted from the discussion because to estimate empirically the effects of price on local calling, measured service data would have to be available. If they were available, then an equation just like equation 20.15 could be constructed for local service. In the absence of these data, however, little can be done, although various attempts to circumvent the problem have been made by using household-level data (see Taylor, 1980).

Estimated versions of equations 21.13 and 21.15 will provide a basis for studying the effects of changes in the prices of telephone services on the proportion of the population subscribing to the system and on the behavior of those who do subscribe. In particular, given an increase in, say, customer line charges required to access the system (*r*), equation 21.13 could be used to predict the change in the number of subscribers to the system. Equation 21.15 could be used to predict the percentage change in the quantity of calls made, given a change in *r*, and thus could serve as a basis for predicting changes in revenues received by the telecommunications carriers.

Business Demand for Telephone Services

Although residential demand for telephone services has been the subject of numerous studies, the demand by business service is a topic that has not been given as much attention. In theory, the problem is straightforward. Just like the demand for any other input into the production process, the demand for telephone services is a derived demand. It is based on the demand for the product being produced by a particular firm or industry and the contribution of the input to the production of the output. In fact, the demand equation for business telephone services—say, by a particular sector of the state economy—might look just like equation 21.14, except that *p* would be interpreted as the price of other inputs, most probably the sectoral wage rate. The variables *N* and *Y* are also included, but they are used to measure changes in the demand for sectoral output. The coefficient signs should remain the same as for the residential demand equations, given these reinterpretations.

To estimate such an equation, data on telephone usage by a number of firms in each industry would have to be available for a period of years. Data on the explanatory variables must also be compiled. As for residential demand, a separate equation should be estimated for each class of service, and the effect of length of haul should be incorporated as in equation 21.10. The product of these efforts will be an estimated equation that can be used to predict the percentage changes in telephone usage, by class, for each business sector of the economy and that will serve as a basis for forecasting the revenue effects of changes in the explanatory variables.

Integrating the Results from Demand Equations into a State Macromodel

The task of integrating the changes in the demands for telephone services by residential and business customers into state-level macroeconometric models as they currently exist is a difficult one. Given estimated versions of the models described earlier, we can predict, for given values of the explanatory variables,

1. The changes in the proportion of the residential population that subscribes to the telephone network.
2. The changes in telephone usage by residential customers for intra-LATA, inter-LATA, and interstate calls.
3. The changes in telephone usage by business customers for intra-LATA, inter-LATA, and interstate calls.
4. The changes in the revenues that will be earned by the telecommunications industry from business and residential customers from intra-LATA, inter-LATA, and interstate calls.

All of these effects must be quantified if we are to assess the total economic impact of changes in telecommunications prices. Obviously, item 4 summarizes items 1 through 3. For product-oriented state models, the changes in the revenues of the telecommunications sector can be input directly, as an autonomous shift in the appropriate intercept (or constant) value, and their effects can be traced through the model. For income-oriented models, the task is somewhat more complicated, since the increase in sector output must act through the demand for labor in the telecommunications sector, and the driving variable in these equations is usually state or national income, not sector output, Thus, a shift in the equation's intercept term must be imputed to capture this effect.

In addition to the effects that we can quantify reasonably well, there are three additional effects for which the direction of change is known but the magnitude is unknown. First, there is the change in the cost of living that occurs because of the changes in telephone prices. The result of this change is, of course, a change in real income for all consumers. In both product- and income-oriented models, this will have the effect of shifting the demand for the output of each sector and, thus, the demand for labor. Therefore, in the demand-for-labor equations, the real income variables will change because of the change in the consumer price index used as a deflator. Consumer price indexes do not exist at the state level, but they are available for the major metropolitan areas, so these values could be used as income deflators. The alternative is to use values of the total U.S. consumer price index (CPI) or gross national product deflator. Some adjustment to the values of these deflators must be made on the basis of the best information available.

Second are the changes in the rate of production in each sector due to the changes in the costs of production associated with modified telephone rates. That is, the supply curve in each sector will shift because of the changes in the cost of production. Since sectoral cost deflators do not exist at the state level, these effects must also be captured, using whatever information is available. For product-oriented models, the intercept term in each sector can be changed by an amount consistent with our knowledge of the cost structure. For income-oriented models, it is the intercept terms in the demand for labor equations that must be altered to reflect the change in the real value of the sector output.

Third are the changes in the use of telephone services relative to other inputs due to the change in their relative prices. This type of change is one that will take some time to occur. Ideally, to measure these effects, the labor demand equations in both product- and income-oriented models would have to be altered to include the price of telecommunications services. This would require a reestimation of the equations for each sector. Then, and only then, can changes in factor usage due to changes in telephone prices be accounted for. However, if there were prior information about input substitutability, and if sensitivities to cost changes were available, the adjustments could be approximated using an add factor on existing equations.

The final step is to combine all these changes with the changes in the U.S. economic variables that are exogenous to the state. Then the state model can be used to generate forecasts of variables that are endogenous to it.

Macroeconomic Models in a Selected Southern Region

One of the primary questions addressed in this analysis is the status of state macroeconomic forecasting models in the selected southern region. In addition, we undertook an assessment of whether or not these models can be appropriately modified to assess the economic effects of changes in carrier access and subscriber line charges. The purpose of this section of the chapter is to summarize the results of those efforts.

The study encompassed nine southern states. At least one letter was written to each modeling group; in most cases, there were several follow-up letters and telephone calls. The response to our inquiries was generally positive, and the individuals exhibited a spirit of cooperation. In some cases, the state forecasting models are up and running and are used routinely by state government agencies in planning for state revenues and expenditure programs. In other cases, the models are being revised and the equations are being reestimated. In one case (Kentucky), the state forecasting model is no longer maintained, and in Alabama, the response to our inquiries was decidedly negative. Thus, there are now seven state forecasting models that could be utilized in assessing the economic impact of changing telecommunications prices. The more specific results from each state are presented in the following pages.

Alabama. The response from Alabama was negative in that they indicated that present commitments precluded their working on a study such as the one proposed here.

Florida. Through repeated correspondence and telephone conversations, it was determined that the Florida models are undergoing extensive revision to include both quarterly and annual models. In addition, models are also being developed for the major metropolitan areas. All of the Florida modeling is based on the income approach. The proposed models are disaggregated, for the most part to the two-digit SIC level. There are separate blocks for (1) transportation and (2) telecommunications and public utilities. Forecasts are made two years into the future in the quarterly model and ten years in the annual model.

Florida has an active and highly competent group of modelers, and it would be possible to proceed rapidly with a study like the one suggested here. This would be one of the better groups to work with in developing a telecommunications subroutine.

Georgia. The Georgia model is undergoing revision under a $50,000 grant from Chase Econometric Associates. The model is quarterly and will be driven by the quarterly Chase macroeconomic model. The proposed Georgia model is based on the income approach, disaggregated to the two-digit SIC level. The part that would have to be revised most extensively is the "Transportation, Telecommunications and Public Utilities" block. Forecasts are made two years into the future. Because of the relationship with Chase, there may be some proprietary restrictions on the use of the model.

Kentucky. We are informed that Kentucky no longer has an operative econometric forecasting model. Before 1981, they had apparently financed a modeling effort with soft money, but during the budget crunch of 1981, they lost the position and no longer maintain the model.

Louisiana. Fairly well developed and operative quarterly and annual models are currently in use in Louisiana. Both models are based on the income approach, and the level of disaggregation is the two-digit SIC level. The block that would have to be revised most extensively is "Transportation, Telecommunications and Public Utilities." Forecasts are made ten quarters and ten years in the future, respectively, for the quarterly and annual models.

Mississippi. The Mississippi model is unique in two respects. First, it is maintained not by a major university but by the Mississippi Research and Development Center. Second, it is the only model in the study that is based on the product approach. The model is annual, and the level of disaggregation is, for the most part, the two-digit SIC level. The model is well developed, operative, and could be used to make economic impact assessments associated with changes in telecommunications prices. The block closely associated with telecommunications is "Transportation, Telecommunications, and Public Utilities." Forecasts are available ten years into the future.

North Carolina. The North Carolina forecasting project is sponsored jointly by First Union National Bank and the University of North Carolina (UNC) at Charlotte. The model is based on the income approach, with the level of disaggregation at the two-digit SIC level. The telecommunications block is embedded in the SIC industry, "Transportation, Telecommunications and Public Utilities." Forecasts are made routinely two years into the future. The model is operative, and the output from the model is available. However, there is no information available about the structural equations or relations.

It may be difficult to obtain much detail about this model, because First Union National Bank apparently has a semiproprietary interest; at least, the UNC-Charlotte group is sensitive to the wishes of First Union, because, among other reasons, First Union provided funding to develop the model.

South Carolina. South Carolina has a well-developed and operative forecasting model that could be used to make economic impact assessments of changes in telecommunications prices. Both quarterly and annual models are available, with forecasts ten quarters or ten years into the future. The level of disaggregation is the two-digit SIC level. Both models are based on the income approach.

Tennessee. Tennessee has an active and mature forecasting model that could be used to make economic impact assessments of changes in telecommunication prices. Both quarterly and annual models will be available for developing scenario analyses. The models are disaggregated, for the most part to the two-digit SIC level. The revised models are based on the income approach.

Directions for Further Study

Alternative Approaches

The Top-Down Approach. This is a standard and elementary approach that would involve taking output from the Wharton study, which is at the national level, and entering those values into the state model as exogeneous variables. The state model would then yield forecasts that could be compared with a state-level base case. This would allow one to assess the state-level effects of telecommunications price changes that originate at the federal level. However, this approach would not be suitable for assessing state-level effects of price changes that emanate from changes in state regulations or that originate at the state level.

It would be possible to adopt this approach even if the state model were not driven by the Wharton model, because all of the state models take national variables in one form or another as exogenous to the system. Of course, the results from the state models would be no more reliable than those of the Wharton model. In fact, they would be even less reliable, because the state-specific effects on wage rates and other variables would have to be input into the model.

The Bottom-Up Approach. This alternative would involve either the reestimation of the endogenous equations with telecommunications prices included in the model or use of the add factor process, whereby an adjustment is made to the intercept to reflect demand shifts in response to shifts in telecommunications prices. However, either procedure would require sector-specific information before one could claim any degree of reliability for the results, and that would be a major effort.

Evaluation of the Alternative Approaches

Suppose that it was decided to proceed with a study of the impact of changes in common carrier access and customer line charges at the state level. What would be an optimal course to pursue and what level of effort would be required?

The Top-Down Approach. The first alternative could be used for any of the state models listed in the preceding section. The procedure would be straightforward, and a relatively low level of effort would be required. The procedure would essentially involve matching the exogenous macroeconomic variables from the Wharton study to the corresponding variables in the state model. Then the state model could be run and the results could be compared with the base case, which most states generate on a routine basis.

If this course were adopted, the best approach would be to negotiate a contract with the state modeling group. Depending on the amount of spade work required of the modelers, and how many of the exogenous input variables were provided to them by outside researchers, the price tag should be relatively low. However, we do not recommend this approach, because the results of such a study would be extremely difficult to defend.

The Bottom-Up Approach. The second alternative would be more satisfactory but also more complicated, since it would involve the specification of a telecommunications subroutine that would become a part of the state model. The telecommunications subroutine would be a block similar to the employment or tax block of the extant model. Such a specification would involve the estimation of a system of equations similar to those specified earlier in this chapter. This would involve specifying the general functional equations, collecting and analyzing the data, specifying the functional form of the equations, and estimating the equations. As a first step, one would need to supplement the existing state-level data base with information about telecommunications usage and expenditures in each sector. It would be a major task, involving perhaps a person-year of effort.

A rational approach would be to negotiate with the state modeling group for the development of a telecommunications subroutine that would be attached to the particular state model. One of the better states would be Florida, because of its size and importance, but there are other good alternatives, including Georgia and Louisiana.

Alternative Avenues of Study

This generic study of economic impact assessments using macroeconomic models in the southern region could be extended in two different directions. One way would be to ascertain the state-specific elasticities of residential and business demand for toll and local service. If these were already available routinely from existing sources, it would simplify matters considerably. On the other hand, if the elasticities must be measured on a state-specific basis, the task of compiling and analyzing the data would be formidable.

The task would be difficult enough for residential subscribers, but for business customers the problems are even more substantial and interesting. In addition to the problems of obtaining a data base suitable for analysis at the industry level, which is highly recommended because of the substantial interindustry differences in the use of telecommunications inputs, the economic theory is less specific. That is, although the theory for input demand is well established, considerable detail at the industry level must be known (for example, the relevant markets, inputs, best measures of economic activity, and so on) to implement that theory at the industry level. In other words, more conceptual work would be required than would be necessary for the residential demand equations. This area of study is relatively undeveloped, and work here would represent a major contribution to the literature.

The second direction would be the further development of a telecommunications subroutine attached to a specific state model. This would offer unique and important challenges to state and regional modeling. Although it is a relatively uncomplicated matter to use existing state models to predict the state-specific impacts of given changes in economic activity at the U.S. level, this kind of forecasting exercise ignores the effects of changes in telecommunications prices at the state level. Consequently, important intrastate and interindustry effects are completely missed.

What is needed is a reconsideration of the structure of state models to take into account, systematically, the effects of major changes in input prices (in this case, telecommunications prices, although the problem is a general one). Although it is generally clear what factors must be considered initially, the specific modeling changes depend on the specific state and state model under consideration. However, given the availability and importance of state modeling efforts, it is clear that efforts in these areas could have a substantial payoff.

Notes

1. The average own-price elasticities for total service demand based on the estimates presented by Taylor (1980) are as follows:

	Residential	*Business*
Local service	−0.10	−0.09
Long-distance service	−1.31	−1.17

2. An example of this general approach is contained in Preston (1975). This is a macroeconomic input-output model that provides links between the demand and production sides of the U.S. economy.

3. The *exogenous variables* are not explained by the model but rather are determined by forces outside the model. They determine the endogeneous variables but are not influenced by them. In other words, the variables of the exogenous variable are assumed to be known and are taken as given for the purpose of the model.

4. The *endogenous variables* are those that are explained by the models. They both determine other variables in the model and are, in turn, determined by other variables.

References

Adams, F. Gerard; Brooking, Carl G.; and Glickman, Norman J. "On the Specification and Simulation of a Regional Econometric Model: A Model of Mississippi." *Review of Economics and Statistics* 57(1975):286–99.

Federal Communications Commission. *Decision and Order in the Matter of MTS and WATS Market Structure,* CC Docket No. 78-72, and *Amendment of Part 67 of the Commission's Rules and Establishment of a Joint Board,* CC Docket 80-286, in CC Docket 84-637, December 19, 1984.

Fomby, T.B.; Hill, R.C.; and Johnson, S.R. *Advanced Econometric Methods.* New York: Springer-Verlag, 1984.

Judge, G.G.; Griffiths, W.E.; Hill, R.C.; Lutkepohl, H.; and Lee, T.C. *The Theory and Practice of Econometrics,* 2nd ed. New York: Wiley, 1985.

Judge, G.G.; Hill, R.C.; Griffiths, W.E.; Lutkepohl, H; and Lee, T.C. *Introduction to the Theory and Practice of Econometrics.* New York: Wiley, 1982.

Latham, William R.; Lewis, Kenneth A.; and Landon, John H. "Regional Econometric Models: Specification of a Quarterly Alternative for Small Regions." *Journal of Regional Science* 19(1979):1–13.

Preston, R.S. "The Wharton Long Term Model: Input Output within the Context of a Macro Forecasting Model." *International Economic Review* 16(February 1975):3–19.

Taylor, Lestor D. *Telecommunications Demand: A Survey and Critique.* Cambridge, Mass.: Ballinger, 1980.

Wharton Econometric Forecasting Associates, Inc. *Impact of the FCC Access Charge Plan on the U.S. Economy.* Philadelphia, Penn.: The Associates (November 1983).

Wharton Econometric Forecasting Associates, Inc. *Background Data and Analysis to Impact of the FCC Access Charge Plan on the U.S. Economy.* Philadelphia, Penn.: The Associates (January 1984).

Index

About the Editors

Albert L. Danielsen is Professor of Economics at the University of Georgia. He received the B.S. from Clemson University in 1960 and the Ph.D. from Duke University in 1966. He recently authored the book *The Evolution of OPEC* and has written numerous articles and research reports on energy and utility markets. Dr. Danielsen has also served in the U.S. Department of Energy as Director of International Market Analysis and Special Assistant and Deputy Assistant Secretary for International Energy Research. He has co-edited a previous volume with David R. Kamerschen, also published by Lexington Books.

David R. Kamerschen is Distinguished Professor and holder of the Chair of Public Utilities in the Department of Economics at the University of Georgia. He received the B.S. and M.A. in 1959 and 1960, respectively, from Miami University and the Ph.D. in 1964 from Michigan State University. He has served as a consultant in utility matters and has presented testimony before numerous state and federal public service commissions. Dr. Kamerschen served as an appointed member to the advisory committee to the Consumers' Utility Counsel for the state of Georgia. He has written more than 130 professional articles and has authored ten books. Dr. Kamerschen also has appeared on public television as a panelist on regulatory economics.